2025 International Conference on Advanced Energy Systems and Power Electronics (AESPE 2025)

Hangzhou, China
13-15 June 2025

IEEE Catalog Number: CFP250H5-POD
ISBN: 979-8-3315-2490-6

**Copyright © 2025 by the Institute of Electrical and Electronics Engineers, Inc.
All Rights Reserved**

Copyright and Reprint Permissions: Abstracting is permitted with credit to the source. Libraries are permitted to photocopy beyond the limit of U.S. copyright law for private use of patrons those articles in this volume that carry a code at the bottom of the first page, provided the per-copy fee indicated in the code is paid through Copyright Clearance Center, 222 Rosewood Drive, Danvers, MA 01923.

For other copying, reprint or republication permission, write to IEEE Copyrights Manager, IEEE Service Center, 445 Hoes Lane, Piscataway, NJ 08854. All rights reserved.

*** This is a print representation of what appears in the IEEE Digital Library. Some format issues inherent in the e-media version may also appear in this print version.

IEEE Catalog Number: CFP250H5-POD
ISBN (Print-On-Demand):   979-8-3315-2490-6
ISBN (Online):   979-8-3315-2489-0

Additional Copies of This Publication Are Available From:

Curran Associates, Inc
57 Morehouse Lane
Red Hook, NY 12571 USA
Phone: (845) 758-0400
Fax: (845) 758-2633
E-mail: curran@proceedings.com
Web: www.proceedings.com

Table of Contents

Simulation Study on the Electromagnetic Shielding Effectiveness of the Shielding Body of Radio Frequency Driven Neutron Generators Xin Tao, Zhi Dou, Yuzhong Qian, Zhenqing Zhu, Mengmeng Li, Lizheng Liang	1
Ultrasonic Excitation Frequency Optimization for Insulation Fault Detection in XLPE Cable based on Sound Field Simulations Wen Lu, Xiaoyu Wang	7
Parameter identification and SOC estimation of energy storage batteries based on HFA-AEKF Longsheng Hua	12
AI Optimization for Coordinated Scheduling of Distributed Energy Storage and Renewable Energy at Multiple Time Scales Junkai Wu	17
A Method for Medium Voltage Underground Cables Incipient fault Location Based on the Sheath Grounding Current Hengzhi Ye, Jinfeng Hu, Jian Dai, Lei Su, Jun Qin, Yuzhe Li, Wenhai Zhang	22
Research on Low-Inductance Packaging Design and Switching Characteristics of GaN HEMT Half-Bridge Modules for High-Frequency Converters Xiangqi Qiu, Xiao Zhang, Pengfei Lu, Song Wei	29
Particle Swarm Optimization-Based Microgrid Dispatch with Renewable Energy Integration Qianlong Li	34
A Wavelet-Enhanced Diagnostic Framework for Internal Short Circuit Detection in Lithium-Ion Batteries Using EIS and XGBoost Jilong Song, Jinshuo Fu, Songtao Che, Hongsi Shi, Kai Wang	39
Multi-distributed energy synergistic regulation method considering demand response Chaoyang Zhi	44
Design of a WPT System in Implantable Medical Devices with Bluetooth-Assisted Alignment Ziluo Ma, Ran Ren, Wenying Zhang	49
Research on Control Strategy of Single-Phase Inverter System Based on Osprey Optimization Algorithm Huili Kang, Wensheng Yan, Shengjie Ma, Lei Weng, Zhen Zhang	54
Open-Circuit-Fault Diagnosis and Fault-Tolerant Strategy for Flying Capacitor Dual-Active-Bridge DC-DC Converter Shunjie Jiang, Wusheng Shi, Dalong Hu, Yuxin Zhou, Wen Guo	58
Research on Cooperative Operation Mode of Pumped Storage Cluster for Regional Power Grid Jiaxing Wang, Chao Huo, Yuxuan Yang, Xiuting Rong, Zun Guo, Zhi An, Ming Zhou, Zhaoyuan Wu	66
Control of DFIG-Based Offshore Wind Farm with Hydrogen Production Connected to DRU-HVDC Zheng Li, Junyang Zhang, Cheng Peng, Yijing Chen, Lei Ba, Chunhua Li, Han Wu, Tao Wang	72
Thermal-Hydraulic Analysis of 7-Rod Bundle Fuel Assembly in Lead-Cooled Fast Reactor Tianyu Hui, Hong Ying, Tao Liu, Qiang Wang, Kewei Fang, Binfeng Wang, Tang Tang, Nan Lin	79
Hybrid Solution Method for Parallel Multi-Channel Flow Allocation Using Quasi-Newton Method and Quantum Genetic Algorithm Tao Liu, Hong Ying, Tianyu Hui, Qiang Wang, Tao Zhang, Haining Shi, Kewei Fang	85
Evaluation of input and output of costly project resources in power grid enterprises System Research Guocheng Li, Yuansheng Duan, Xiaoxue Ma, Can Tao	91
Research on Energy Storage Mechanism of Polymer Composite Materials Based on Interface Structure Regulation Chen Chen, Lifang Shen, Shubin Yan, Guang Liu and Yang Cui	96
Application Research of Digitalization of Flow Batteries in New Power Systems Fuquan Hu, Xiaoqiang Liu, Xiaowei Wang, Lin Yang	101
A Nash-Stackelberg Game Equilibrium Approach for Electricity Markets Using RBF-Based Value Function	105

Approximation Yun Yang, Kai Li, Yue Zhao, Minjing Yang, Zichao Meng, Yuhao Luo, Jianquan Zhu	
Optimizing Power Decarbonization Pathways: A Scenario-Based Portfolio Planning Approach Haifeng Qiu, Sichao Chen, Liguo Weng, Hui Pan, Lingzhen Shi	111
Simulation of slurry heat transfer in heat exchanger tube Jie Dou	118
Research on Reasonable Injection Production Ratio in Different Water Content Stages of Block S Lingling Chen	124
Geological Conditions and Hydrocarbon Accumulation Mechanism Analysis of Tight Oil Reservoirs in Daqing Oilfield Jing Chen	131
Enhanced Multi-Frequency Virtual Oscillator Control for Simultaneous Harmonic Suppression in Hybrid Alternating Current/Direct Current Microgrids Zhihao Zhang, Zongliang Wang, Wangxu Chu, Weixun Wu, Jiaming Wen	136
Research on the Application of Artificial Intelligence in Long-Distance Pipeline Leakage Monitoring and Risk Prediction Zeji Li, Zhanyi Xue	140
Author Index	145

Preface

The 2025 International Conference on Advanced Energy Systems and Power Electronics (AESPE2025) is hosted by Hangzhou Dianzi University and will take place from June 13 to 15, 2025, in Hangzhou, China.

AESPE2025 is a premier academic event dedicated to exploring the latest research findings and innovative advancements in advanced energy systems and power electronics technology, which provides a warm and welcoming platform for leading experts, researchers, and practitioners to share insights and foster collaborations in this rapidly evolving field.

The conference proceedings collect papers related to Advanced Energy Systems (including Renewable Energy Technologies and Systems, Renewable Energy Storage Technology and Nuclear Energy, etc.) and Power Electronics (Power Electronic Circuits, Smart Grids and Microgrids and Power Electronics and Applications, etc.). All papers are reviewed by two or three expert reviewers and selected based on originality, relevance, technical significance, completeness, and writing structure.

AESPE2025 invites many scholars from home and abroad to report and discuss their innovative findings in the fields, including Prof. Jahangir Hossain (from University of Technology Sydney, Australia, IEEE senior member) as keynote speaker to share his knowledge in renewable energy and energy storage systems domains.

The Organizing Committee of AESPE2025 sincerely express our gratitude to all the participants, keynote speakers, editors, authors, committee members, volunteers, staff and organizers for their contributions and precious time. We look forward to meet you in the next edition and have academic exchanges together.

The Organizing Committee of AESPE2025

Committee Member

Conference General Chairs
Prof. Liang Chu, Hangzhou Dianzi University, China
Prof. Guijie Liang, Hubei Key Laboratory of Low-Dimensional Optoelectronic Materials and Devices, Hubei University of Arts and Sciences, China

Publication Chairs
Assoc. Prof. Zang Yue, Hangzhou Dianzi University, China
Prof. Mahdi Deymi Dashtebayaz, Hakim Sabzevari University, Iran

Organizing Committee Chairs
Prof. Wangnan Li, Hubei Key Laboratory of Low-Dimensional Optoelectronic Materials and Devices, Hubei University of Arts and Sciences, China
Prof. Shuting Pang, Hangzhou Dianzi University, China

Technical Program Committee Chairs
Prof. Fushuan Wen, Zhejiang University, China
Assoc. Prof. Qin Zhou, Hangzhou Dianzi University, China

Organizing Committee Members
Prof. Xiangjie Kong, Zhejiang University of Technology, China
Prof. Kaikai Chi, Zhejiang University of Technology, China
Assoc. Prof. Kechen Zheng, Zhejiang University of Technology, China
Assoc. Prof. Xiaoying Liu, Zhejiang University of Technology, China
Assoc. Prof. Guodong Li, Hangzhou Dianzi University, China
Assoc. Prof. Jingya Wei, Hangzhou Dianzi University, China
Assoc. Prof. Norzanah Rosmin, Universiti Teknologi Malaysia, Malaysia
Assoc. Prof. Aripriharta, Universitas Malang, Indonesia
Assoc. Prof. Rasul Mohebbi, Damghan University, Damghan, Iran
Assoc. Prof. Ali Ayati, Quchan University of Technology, Iran
Assoc. Prof. Andrey Nikitin, ITMO University, Russia
Dr. Hamid Reza Rahbari, University of Denmark, Denmark
Dr. Xiao Xu, Hangzhou Dianzi University, China
Dr. Peng Zhou, Hangzhou Dianzi University, China
Dr. Yu Wang, Hangzhou Dianzi University, China

Technical Program Committee Members
Prof. A. Pandian, Koneru Lakshmaiah Education Foundation (Deemed to be University), India
Assoc. Prof. Arvind R. Singh, Hanjiang Normal University, China
Assoc. Prof. Wasim Abbas, Hanjiang Normal University, China
Assoc. Prof. Surender Reddy Salkut, Woosong University, South Korea
Asst. Prof. Rajaram Tukaram Ugale, COEP Technological University, India
Asst. Prof. Bhupendra Kumar, G H raisoni Institute of Engineering and Technology, India
Asst. Prof. Pradeep Vishnuram, SRM Institute of Science and Technology, India

Simulation Study on the Electromagnetic Shielding Effectiveness of the Shielding Body of Radio Frequency Driven Neutron Generators

XinTAO[1,a]
[1]Institute of Energy, Hefei Comprehensive National Science Center (Anhui Energy Laboratory)
Anhui, Hefei, China
e-mail:taoxin@ie.ah.cn

YuZhong QIAN[1,c*]
[1]Institute of Energy, Hefei Comprehensive National Science Center (Anhui Energy Laboratory)
Anhui, Hefei, China
*[c] Corresponding author e-mail:yzqian@ie.ah.cn

MengMeng LI[2,e]
[2]Anhui University of Science and Technology
Anhui, Huainan, China
[e]e-mail: 2179555200@qq.com

Zhi DOU[2,b]
[2]Anhui University of Science and Technology
Anhui, Huainan, China
[b]e-mail: 306837639@qq.com

ZhenQing ZHU[2,d]
[2]Anhui University of Science and Technology
Anhui, Huainan, China
[d]e-mail: 2964523079@qq.com

LiZheng LIANG[1,3,f]
[1]Institute of Energy, Hefei Comprehensive National Science Center (Anhui Energy Laboratory)
[3]Institute of Plasma Physics, Chinese Academy of Sciences
Anhui, Hefei, China
[f]e-mail: lzliang@ipp.ac.cn

Abstract—**During the operation of radio frequency driven neutron generators, the electromagnetic energy generated by the radio frequency inductively coupled coil spreads to the surrounding area, causing severe radio frequency interference to the communication system. This can lead to a loss of control of the communication system, significantly affecting the experimental process and posing considerable safety risks. Therefore, it is necessary to implement electromagnetic shielding for these generators. A model of the radio frequency inductively coupled coil and its shielding body was established. To balance electromagnetic shielding and heat dissipation, circular holes were created on the surface of the shielding body, and the impact of different opening rates on the electromagnetic shielding effectiveness of the shielding body was analyzed. The simulation results indicate that the electromagnetic shielding effectiveness of the shielding body decreases with the increase of the opening rate. When the opening rate is 15.8%, the change in the inductance value of the radio frequency inductively coupled coil is minimal, with only a 5.7% variation, which has the least impact on the power coupling of the radio frequency system. The analysis also revealed that the electric field strength at the center position of each face of the shielding body is higher than that at the ends, with a maximum value approximately 1.74 times greater than the electric field strength at the end positions. Under the same total opening rate, shielding bodies with smaller hole radius and greater thickness exhibit higher shielding effectiveness. The simulation results provide valuable reference for the electromagnetic shielding design of radio frequency driven neutron generator devices.**

Keywords- Electromagnetic shielding of radio frequency driven neutron generators; total opening rate; opening position; opening radius; shielding body thickness

I. INTRODUCTION

Neutrons are electrically neutral and have strong penetrating power, making them ideal probes for studying the microscopic structure and properties of matter, as they can distinguish between light elements, isotopes, and neighboring elements. With the development of neutron technology, neutrons have been widely applied in various fields such as life sciences, national security, geochemistry, environmental science, archaeology, and materials science[1]. Accelerator neutron generators, as important scientific devices for neutron production, research, and utilization, also serve as key experimental platforms for the development and application of neutron technologies. In recent years, significant research progress has been made both domestically and internationally.

Radio Frequency Inductive Coupling (RF-ICP) plasma sources, with their advantages of high ion density, compact design, and stable long-pulse operation, can serve as the ion source for accelerator neutron generators. The working principle of the Radio Frequency Inductive Coupling plasma source is as follows: The radio frequency power source transmits RF power to the coupling coil through an impedance matching system. When RF current flows through the coupling coil, it generates an RF magnetic field around the discharge chamber. Under the influence of the RF magnetic field, electrons in the gas to be ionized gain a significant amount of energy and collide with gas molecules, thereby producing plasma. However, in practical applications, due to the use of radio frequency excitation, the electromagnetic energy produced by the RF coupling coil not only couples into the plasma but also diffuses into the surrounding environment,

979-8-3315-2490-6/25 $31.00 © 2025 IEEE

leading to the loss of control of the communication system in the RF-driven neutron generator. This has a particularly severe impact on the tower-type high-voltage power supply for the accelerator tube section, as its tower design structure is more susceptible to electromagnetic interference from the RF coil, resulting in the loss of communication in the high-voltage power supply. This significantly affects the progress of experiments and poses considerable safety risks. Therefore, it is necessary to address the electromagnetic compatibility issues of RF-driven neutron generators to ensure their safe and reliable operation.

II. ESTABLISHMENT OF THE SHIELDING BODY SIMULATION MODEL AND PARAMETER SETTINGS

A. Definition of Shielding Effectiveness

To address the electromagnetic compatibility issues present in RF-driven neutron generators, a shielding body can be added around the RF inductive coupling coil to isolate or reduce the propagation of electromagnetic energy. The shielding body can be made of any conductive and magnetic material, and its primary function is to confine the electromagnetic field generated by the RF inductive coupling coil as much as possible within the boundaries of the shielding body, thereby reducing the diffusion of electromagnetic energy to the outside and minimizing interference with other communication systems[2].

The definition of shielding effectiveness is a measure of the attenuation of electromagnetic waves by a shielding body, usually expressed in decibels (dB). A higher value indicates a stronger shielding effect of the shielding material. The formula for calculating shielding effectiveness is as follows[3]:

$$\begin{cases} SE = 20\log_{10}\left(\dfrac{E_1}{E_2}\right) \\ SE = 20\log_{10}\left(\dfrac{H_1}{H_2}\right) \end{cases}, \qquad (1)$$

where the E_1 (or H_1) represents the magnitude of the electric field strength (or magnetic field strength, respectively.) at a certain point in space without the shielding body, while E_2 (or H_2) represents the magnitude of the electric field strength (or magnetic field strength) at that point after the shielding body has been placed.

B. Construction of the Simulation Model

Due to the need to shield the electromagnetic energy generated by the RF coupling coil to prevent interference with the communication system of the RF-driven neutron generator, a simulation analysis using internal radiation source shielding was adopted[4]. Specifically, the RF coupling coil is placed inside the shielding body as the radiation source, while an electric field probe is positioned outside the shielding body to monitor the electric field strength at that point.

First, the RF coupling coil model is constructed as shown in Figure 1. The coil material is selected as copper (annealed) as the coil material, with an electrical conductivity of: 5.8×10^8

S/m, and a permeability of 1, with a wire radius of 1.5 mm, a pitch of 3.5 mm, a total of 4.5 turns, and a coil radius of 15 mm. The length of the wire at both ends of the coil is 100 mm. A grounding plane is set at the lower end of the coil (The material selected for the grounding plane is the same as that of the coil, and its thickness is set to 0, which is equivalent to an ideal conductor that prevents electromagnetic signals from penetrating. Theoretically, the grounding plane serves as a good shielding body. However, in this study, the grounding plane and the coil model are considered as an integrated internal radiation source, and subsequent simulation studies modify the parameters of the external shielding body based on this model. Therefore, the grounding plane is not included in the shielding range of the shielding body), and the midpoints of the two wires are selected along with their corresponding points mapped to the grounding plane. The excitation signal is defined using a discrete port method, the boundary of the simulation model is set to open (and space) in this model.

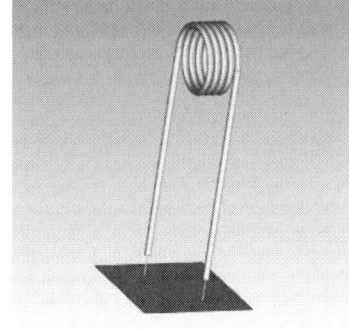

Figure 1.   RF Coupling Coil Model

After constructing the RF coil model, the shielding body model is built, as shown in Figure 2. The thickness of the shielding body is 2 mm, with a length of 80 mm, a width of 75.75 mm, and a height of 150 mm.

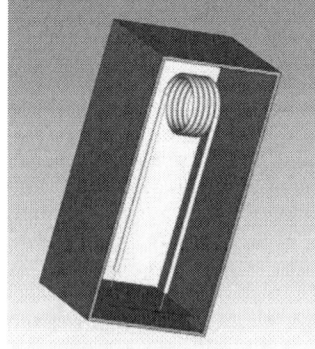

Figure 2.   Cross-section of the Shielding Body Model

Since the RF frequency used in the RF-driven neutron generator system is 13.56 MHz, the simulation frequency range is set from 0 to 100 MHz. An electric field probe is placed 50 mm around the shielding body to measure the electric field strength at that location.

III. SIMULATION RESULTS ANALYSIS

A. Analysis of Shielding Effectiveness with Different Aperture Ratios

During operation, the RF inductive coupling coil generates a significant amount of heat. While full shielding effectively isolates the propagation of electromagnetic energy, it is not conducive to heat dissipation, which may greatly affect the stable operation of the RF inductive coupling ion source. Creating apertures on the surface of the shielding body facilitates heat dissipation; however, these openings can disrupt the continuity of the conductive surface[5], thereby reducing the shielding effectiveness. Existing research indicates that under the same individual aperture area and comparable processing difficulty, circular holes provide better shielding effectiveness than holes of other shapes. This study selects different diameters of circular apertures on the surface of the shielding body to analyze the impact of different aperture ratios on the shielding effectiveness.

As shown in Figure 3. the models of the shielding bodies with aperture ratios of 0%, 15.8%, 34.2%, and 100% are presented. An aperture ratio of 0% indicates full shielding; for the aperture ratio of 15.8%, the radius of the circular hole is 1 mm with a thickness of 2 mm; for the aperture ratio of 34.2%, the radius of the circular hole is 2.5 mm with a thickness of 2 mm; and an aperture ratio of 100% indicates no shielding. The measured electric field strengths for the four aperture ratios are shown in Figure 4. To facilitate the calculation of shielding effectiveness, the unit of the electric field strength has been converted to dB(V/m). From the figure, it can be observed that as the aperture ratio of the shielding body increases, the electric field strength at the measurement points also increases, indicating a decreasing trend in the shielding effectiveness of the shielding body with the rise in aperture ratio. At a frequency of 13.56 MHz, the shielding effectiveness SE (32.4%) of the shielding body with an aperture ratio of 32.4% is 37.53 dB, while the shielding effectiveness SE (15.8%) of the shielding body with an aperture ratio of 15.8% is 58.23 dB. For the shielding body with an aperture ratio of 0%, the measured electric field strength is 0, indicating that all electromagnetic energy is completely shielded by the shielding body.

In the RF-driven neutron generator device, the change in inductance of the RF inductive coupling coil is a key factor affecting the power coupling of the RF system. TABLE I. shows the values of the RF coupling coil inductance at a frequency of 13.56 MHz for four different aperture ratios.

TABLE I. Values of the Imaginary Part of the Impedance of the RF Coil at Four Different Aperture Ratios.

Total aperture ratio (%)	Inductance value(μH)	Inductance variation rate（%）
0	0.652	7.5
15.8	0.665	5.7
32.4	0.657	6.8
100	0.705	/

From TABLE I. it can be seen that when the opening ratio is 15.8%, the variation rate of the inductance value of the RF inductive coupling coil is the smallest, changing only by 5.7%

compared to the unshielded case. This indicates that at an opening ratio of 15.8%, the shielding has the least impact on the power feeding of the RF system.

Based on the simulation analysis results in Figure 4. and TABLE I. and taking into account both the electromagnetic shielding and heat dissipation of the RF-driven neutron generator system, a shielding structure with an aperture ratio of 15.8% is selected for subsequent simulation analysis.

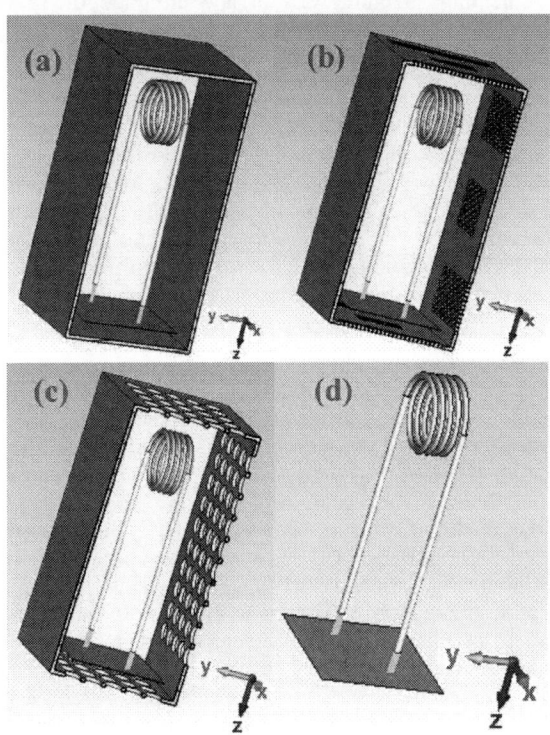

Figure 3. Cross-sectional view of the shielding body model with different opening rates：(a) Model with opening rate of 0% (b) Model with opening rate of 15.8% (c) Model with opening rate of 34.2% (d) Model with opening rate of 100%

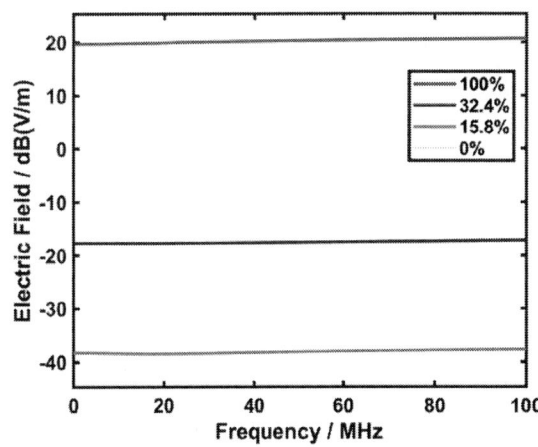

Figure 4. Electric Field Intensity at Aperture Ratios of 0%, 15.8%, 34.2%, and 100%

B. Analysis of Shielding Effectiveness at Different Shielding Positions

From the analysis in the previous subsection, it can be concluded that the shielding structure with an aperture ratio of 15.8% exhibits a good shielding effectiveness. In order to investigate the impact of openings at different positions on the shielding effectiveness of the shielding structure, electric field probes were placed at various locations on the four surfaces of the shielding structure with an aperture ratio of 15.8% to monitor the electric field intensity at those positions.

Figure 5. shows the distribution of probes on the surfaces of the shielding structure. A total of 28 electric field probes were placed: 5 probes on both the top and bottom surfaces of the shielding structure, and 9 probes on each of the side surfaces. These probes were positioned 50 mm away from the surface of the shielding structure to monitor the electric field intensity at different points on each surface. Running the simulation analysis, the electric field intensity curves at different positions on each surface were obtained and are shown in Fig. 6.

As shown in Fig. 6. the electric field strength at the top surface of the shielding structure first decreases and then increases with the rise in frequency. The electric field strength measured by Probe 3, located at the center of the top surface, is the highest. At a frequency of 13.56 MHz, the electric field strength at Probe 3 is approximately 71.4% higher than that at Probes 2 and 4. The electric field strength at the bottom surface of the shielding structure shows a gradually increasing trend with the rise in frequency. Since the bottom surface is close to the grounding surface of the coil, the average electric field strength at the bottom surface is lower than that at the top surface, and the electric field strength at different positions on the bottom surface varies only slightly. At a frequency of 13.56 MHz, the electric field strength measured by Probe 8 at the center position is the highest, approximately 5.8% greater than the electromagnetic field strength at the positions of Probes 7 and 9. The electric field strength at Side Surface 1 of the shielding structure shows a gradually increasing trend with the rise in frequency. Although there are no openings below the positions of Probes 12 and 18, the electric field strength measured by the row of probes (Probes 12, 15, and 18) at the center position of Side Surface 1 is significantly higher than that at the upper and lower ends. Among them, Probe 15 records the highest electric field strength. At a frequency of 13.56 MHz, the electric field strength at the position of Probe 15 is approximately twice that of the electric field strength at the upper and lower ends of Side Surface 1. The electric field strength at Side Surface 2 of the shielding structure shows a trend of first decreasing and then increasing with the rise in frequency. Similar to the results at Side Surface 1, the electric field strength in the middle row is higher than that at the upper and lower ends. Probe 24, located at the center position of Side Surface 2, measures the highest electric field strength, which is approximately 1.74 times greater than that at the upper and lower ends at a frequency of 13.56 MHz.

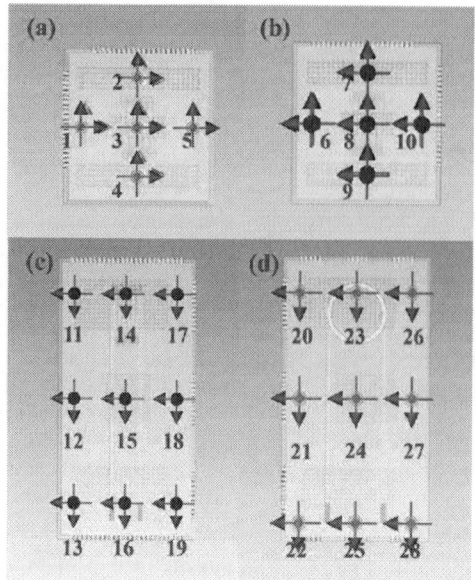

Figure 5. Probe Distribution on the Shielding Structure: (a) Position of the top surface probe (b) Position of the bottom surface probe (c) Position of the side 1 probe (d) Position of the side 2 probe

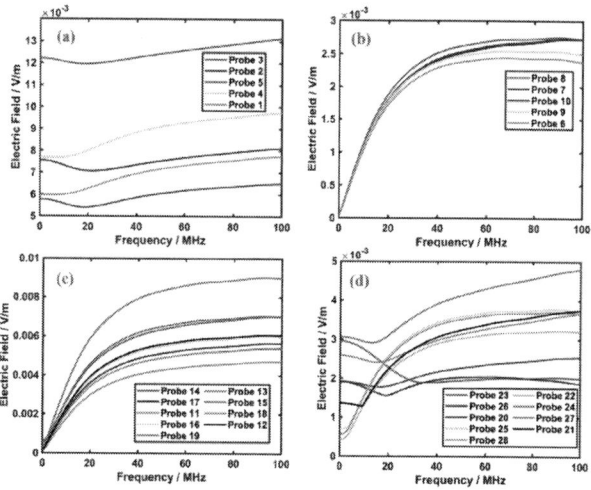

Figure 6. Electric Field Strength at Different Positions on Each Surface: (a) Electric field intensity at different positions on the top surface (b) Electric field intensity at different positions on the bottom surface (c) Electric field intensity at different positions on side 1 (d) Electric field intensity at different positions on side 2

From the above analysis, it can be concluded that when designing electromagnetic shielding for the radio frequency-driven neutron generator system, openings should be avoided at the middle position of the shielding structure. Instead, it is advisable to choose to create openings at the upper and lower ends of each surface to enhance the shielding effectiveness of the structure.

C. Analysis of Shielding Effectiveness with the Same Opening Ratio but Different Opening Diameters

From the above analysis, it can be concluded that as the opening ratio increases, the shielding effectiveness of the shielding structure decreases. However, under the condition of the same total opening ratio, different radius of circular openings may also have an impact on the shielding effectiveness of the structure[6]. In order to analyze the effect of circular hole diameter on the shielding effectiveness of the shielding structure, this section compares the electric field strength on the surface of the shielding structure under the condition of a total opening ratio of 15.8% (with an opening ratio error of ±0.2%) for circular hole radius of 1mm, 3mm, and 5mm. The simulation results are shown in Figure 7. below.

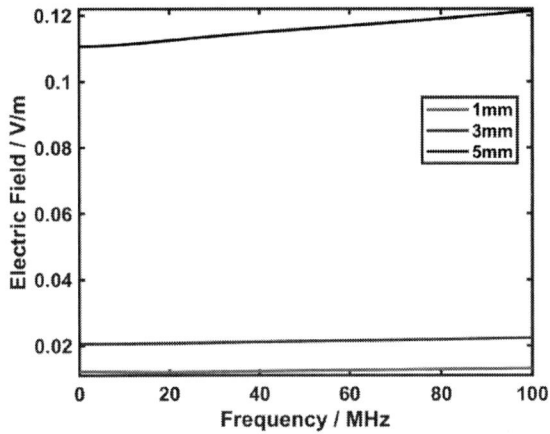

Figure 7. Variation of Electric Field Strength with Frequency for Different Opening Diameters at the Same Total Opening Ratio.

As shown in Figure 7. , the electric field intensity measured under three different opening radius schemes increases with the frequency. The increase in electric field intensity for the circular shields with opening radius of 1 mm and 3 mm is significantly slower compared to the shield with an opening radius of 5 mm. Additionally, as the opening radius of the shield increases, the electric field intensity measured at the probe also increases. At a radio frequency of 13.56 MHz, the electric field intensity measured on the surface of the circular shield with a 1 mm opening radius is the lowest. The electric field intensity measured on the surface of the circular shield with a 3 mm opening radius is approximately twice that of the 1 mm opening radius, while the electric field intensity measured on the surface of the circular shield with a 5 mm opening radius is about ten times that of the 1 mm opening radius. From the above simulation results, it can be concluded that under the same total opening rate, as the opening radius of the circular shield increases, the shielding effectiveness exhibits an exponential decrease..

The reduction in shielding effectiveness may be attributed to the fact that as the opening radius increases, it gradually approaches or even exceeds half the wavelength of the electromagnetic wave ($\lambda/2$). At this point, the circular holes can lead to a sharp increase in the penetration efficiency of the electromagnetic waves, resulting in a significant decrease in shielding effectiveness. From the above results, it can be concluded that when designing electromagnetic shielding for radio frequency-driven neutron generator systems, the opening radius should be strictly controlled to be much smaller than the wavelength of the electromagnetic waves in order to enhance the shielding effectiveness of the shield.

D. Analysis of Shielding Effectiveness for Different Shielding Thicknesses

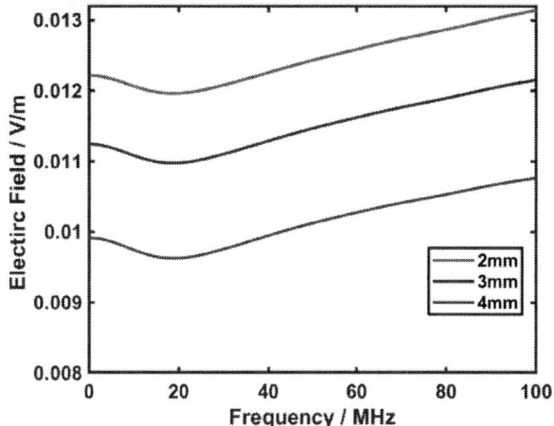

Figure 8. Variation of Electric Field Strength with Frequency for Different Shielding Thicknesses

According to the mechanism of electromagnetic shielding, the thickness of the shielding material is also a key parameter that affects shielding effectiveness[7]. Under the conditions of a total opening ratio of 15.8% and an opening radius of 1mm, the electric field strength measured outside the shielding body for thicknesses of 2mm, 3mm, and 4mm is compared, as shown in Figure 8. In the figure, as the frequency increases, the electric field strength measured outside the shielding bodies of the three different thicknesses shows a trend of initially decreasing and then increasing. At the same time, as the thickness of the shielding body increases, the electric field strength measured outside the shielding body gradually decreases. At a frequency of 13.56 MHz, the electric field strength outside the 4mm thick shielding body is the lowest, approximately 20% lower than that of the 2mm thick shielding body. This indicates that as the thickness increases, the shielding effectiveness of the shielding body improves.

From the mechanism of electromagnetic shielding, the loss of electromagnetic waves after passing through the shield can be mainly divided into three parts: reflection loss (SE_R), absorption loss (SE_A), and multiple reflection factor (SE_M). Their expressions are as follows[7]:

$$SE_T = SE_R + SE_A + SE_M . \qquad (2)$$

According to the theory of electromagnetic fields, SE_R can be expressed by the following formula[8]:

$$SE_R = 20 \lg e^{t/\delta} = 0.868t / \delta = 1.131t\sqrt{fu_r\sigma_r} , \qquad (3)$$

where the δ is the skin depth of the shield (m), μ_r is the magnetic permeability of the shield (H/m), σ_r is the electrical

979-8-3315-2490-6/25 $31.00 © 2025 IEEE

conductivity of the metal plate (S/m), f is the frequency of the electromagnetic wave (Hz), and t is the thickness of the shield (mm).

From Equation (3), it can be observed that the greater the thickness of the shield, the higher the loss due to the reflection of electromagnetic waves when they impinge on the surface of the shield, resulting in better shielding effectiveness. Therefore, in Figure 8, when the thickness of the shield is set to 4 mm, the electric field intensity measured on the surface of the shield is significantly lower than that at 2 mm and 3 mm, indicating the best shielding effectiveness. Thus, when designing electromagnetic shielding for radio frequency-driven neutron generator systems, a thicker shield should be selected to enhance shielding effectiveness while also considering heat dissipation.

IV. CONCLUSIONS

This paper established a simulation model for the radio frequency inductive coupling coil and its electromagnetic shielding box in a radio frequency-driven neutron generator device, and analyzed the shielding effectiveness of the shielding body at opening ratios of 0%, 15.8%, 32.4%, and 100%. By analyzing the simulation results, it was found that the shielding effectiveness of the shielding body gradually decreases as the opening ratio increases. When the opening ratio is 15.8%, the change in the inductance value of the radio frequency inductive coupling coil is minimal, resulting in the least impact on the stable power input of the radio frequency system.

Furthermore, the study investigated the effects of different opening positions on the surface of the shielding body, the same opening ratio, different opening radius, and varying shielding body thicknesses on the electromagnetic shielding effectiveness of the shielding body when the opening ratio is 15.8%. The results show that the electric field strength is highest at the center position of each surface of the shielding body. Under the same opening ratio, a smaller opening radius and a greater shielding body thickness lead to better electromagnetic shielding effectiveness. Therefore, when designing the electromagnetic shielding body for the radio frequency-driven neutron generator device, it is advisable to create openings at the center position of each surface of the shielding body. Additionally, under the same total opening ratio, a shielding body with smaller opening radius and greater thickness will have better shielding effectiveness. This result has certain reference significance for the design of the electromagnetic shielding body in radio frequency-driven neutron generator devices.

ACKNOWLEDGMENT

This research was financially supported by the National Natural Science Foundation of China (12105135, 12305200), and the University Synergy Innovation Program of Anhui Province GXXT-2021-015 and GXXT-2022-005, and the Hefei Comprehensive National Science Center under Grant Nos. 21KZS202 and 21KZS208.

REFERENCES

[1] Xiaohou Bai; Jun Ma; Zhiyuan Wei; Jingying Wang; Xiaoxue Yu; Shiyu Zhang, "Development of a high-yield compact D-D neutron generator," Nuclear Inst. and Methods in Physics Research, A, 2024, 1069 169993-169993.

[2] Truhn, Kiessling, and Schulz, "Optimized RF shielding techniques for simultaneous PET/MR"

[3] Jia Xiao, ZhengXiang Song, Jianhua Wang, Liang Wang,"Simulation Analysis of Electromagnetic Shielding of Electronic Device Chassis," 2019 12th International Workshop on the Electromagnetic Compatibility of Integrated Circuits (EMC Compo), October 21-23, Haining, China.

[4] Xu Lei. Internal and ExTERNAL Radiation Source Methods for Evaluating Electromagnetic Shieldding Effectiveness[J]. SAFETY&EMC,2024(6):54-58..

[5] Güler Sunay. An investigation on electromagnetic shielding effectiveness of metallic enclosure depending on aperture position[J]. Journal of Microwave Power and Electromagnetic Energy, 2023, 57(2):129-145.

[6] Zhao Xinyang, Yin Qiyun, Yang Chen, Lu Hongjian. Influence of aperture size parameters on 5G shielding effectiveness of shielding cavity[J]. Application of Electronic Technique, 2023, 49(4) : 16-20.

[7] CHENG J, LI C, XIONG Y, et al. Recent Advances in Design Strategies and Multifunctionality of Flexible Electromagnetic Interference Shielding Materials [J]. Nano-Micro Letters, 2022, 14(1): 80-111.

[8] LIU Li, lIU Bin, GE Yushi, LIU Weiqun, DING Yong,GUO Yong, "Research of the shielding effectiveness of the metallic enclosure with an aperture" DOI： 10.14022/j.issn1674-6236.2021.13.016

2025 International Conference on Advanced Energy Systems and Power Electronics (AESPE 2025)

Ultrasonic Excitation Frequency Optimization for Insulation Fault Detection in XLPE Cable based on Sound Field Simulations

Wen Lu
College of Intelligent
Transportation, Chongqing
Vocational College of Transportation, Chongqing
Chongqing, 402247, China
luxiya0608@163.com

Xiaoyu Wang
College of Intelligent
Transportation, Chongqing
Vocational College of Transportation Chongqing
Chongqing, 402247, China
cqggyswxy@163.com

Abstract—**This paper, based on a Multiphysics simulation model coupling ultrasonic sensors with the structure of high voltage cables, systematically analyzes the propagation characteristics of ultrasonic guided waves through multi-layer media such as the buffer layer, insulation layer, and conductor shielding layer. The relationship between the ultrasonic guided wave propagation characteristics across the cross-section of the high-voltage cable and the excitation frequency is investigated. The results show that when a low-frequency excitation of 0.5 MHz is used, the ultrasonic energy attenuation is relatively low, providing stronger penetration capability, allowing the waves to pass through the conductor shielding layer and reach the copper core. However, the reflected echo strength is weak, and the time-domain features are unclear. As the excitation frequency increases to 3.0 MHz and 5.0 MHz, the acoustic impedance matching characteristics of the conductor shielding layer improves, leading to a significant enhancement in echo signal strength, and the signal characteristics become clearer. However, the increase in ultrasonic wave attenuation reduces its ability to penetrate the conductor shielding layer, making it difficult for the waves to effectively reach the copper core.**

Keywords—*high-voltage cable; ultrasonic waves; excitation frequency; non-destructive testing;*

I. INTRODUCTION

Power cables as core transmission equipment in modern power systems are experiencing rapid development due to the increasing demand in transformation of the energy structure and the power supply. Especially with the promotion of policies such as the undergrounding of overhead lines in large cities, the application scope and market demand for power cables are continually expanding. However, due to the impact of manufacturing processes and external factors such as mechanical stress and thermal stress during operation, high-voltage cables may suffer from defects such as cracks, voids, and deformations. These flaws can compromise the mechanical strength and sealing integrity of the cables, thereby impacting the overall reliability, and posing a risk to the safe and stable functioning of power systems [1, 2]. Therefore, timely and accurate detection of internal defects in power cables is crucial for ensuring the safety and reliability of the power system. In defect detection, non-destructive testing technology is the primary method due to its ability to assess the cable without damaging its structure. Among them, ultrasonic testing is one of the most popular non-destructive testing methods known for its high sensitivity and good effectiveness [3, 4]. However, the performance of ultrasonic testing is very sensitive to excitation frequencies, which put it a technical challenge for detecting cable internal defects, since the frequency response varies in different layers of cable materials [5]. It is necessary to conduct in-depth research on the propagation characteristics of different excitation frequencies in high-voltage cables, to achieve excitation frequency optimization.

This study uses COMSOL Multiphysics finite element software to conduct purely numerical simulations without involving actual hardware experiments. The excitation frequencies of 500 kHz, 1.5 MHz, 3.0 MHz, and 5.0 MHz were selected to cover typical low to high-frequency ranges commonly used in practical ultrasonic testing equipment, aiming to align the simulation settings with real-world detection scenarios and enhance the applicability of the results. This allows for a systematic analysis of how frequency variations affect ultrasonic propagation characteristics, providing theoretical support for future experimental studies and engineering applications. The simulation results show that low-frequency ultrasonic waves, with longer wavelengths and lower attenuation, can effectively penetrate the deeper structures of the cable, such as the conductor shielding layer and copper core, making them suitable for detecting deep internal defects. On the other hand, high-frequency ultrasonic waves, with shorter wavelengths and stronger attenuation, lose energy quickly and have weaker penetration ability, making them more suitable for detecting surface defects in the cable.

II. MODELING METHODOLOGY

A. Geometric Modeling

This paper takes the internal structure of a standard 110 kV high-voltage cable as the research object and establishes a simulation model of the cable, as shown in Fig. 1. The model uses multiphysics coupling of the solid mechanics module and electrostatics module for simulation and employs the finite element method to study the propagation behavior of ultrasonic waves in complex media. The model accurately reflects the impact of each material layer in the cable on the propagation of ultrasonic waves, providing a reliable simulation basis for high-voltage cable defect detection. Since high-voltage cables are complex, curved structures typically composed of multiple layers of different materials, including aluminum sheathing, outer sheath, and insulation layers, this study simplifies the model by only considering the material layers up to and including the buffer layer for analysis. Therefore, understanding and analyzing the impact of these material layers on ultrasonic wave propagation is crucial for optimizing ultrasonic testing methods for high-voltage cables.

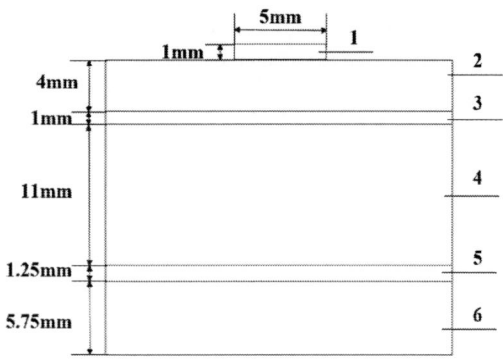

Figure 1. Hierarchical structure of high-voltage cable simulation

This paper uses the actual dimensions of the wrinkled high-voltage cable, with certain simplifications made to the hierarchical structure. Specifically, in this model, the piezoelectric patch of the cable is made of lead zirconate titanate material, with a width of 5.0 mm and a height of 1.0 mm, to simulate its response to ultrasonic waves. The insulating layer and shielding layer are primarily made of cross-linked polyethylene, with thicknesses of 11.0 mm and 1.0 mm, respectively. These materials are used to isolate the electrical components and ensure the mechanical properties of the cable. The conductor of the cable is made of copper, with a thickness of 5.75 mm, to ensure the electrical conductivity of the cable. The water-blocking buffer layer is primarily made of polyethylene, with a thickness of 4.0 mm. The dimensions of the different layers of the cable are shown in Table I, while the key properties of the main components of the cable are listed in Table II.

For the multiphysics modeling, based on the relationship between ultrasonic wavelength and meshing size, to ensure calculation accuracy, each wavelength needs to occupy within 5 to 8 mesh cells. This meshing criterion ensures adequate spatial resolution to accurately resolve wave propagation phenomena, such as reflection, refraction, and scattering, especially in heterogeneous media. Therefore, as the ultrasonic frequency increases, the corresponding wavelength decreases, allowing for more mesh cells to be defined within the same computational domain, resulting in higher spatial resolution. This characteristic gives high-frequency ultrasound a significant advantage in numerical simulations that require fine structures, as it can more accurately capture subtle variations in the sound field and complex structural features, such as small-scale defects, sharp interfaces, or intricate material geometries. In this paper, mesh refinement is conducted for frequencies of 500 kHz, 1.5 MHz, 3.0 MHz, and 5.0 MHz, with the corresponding meshing results presented in Fig. 2.

To provide a clearer illustration of the modeling and simulation process in this study, a simulation flowchart is presented in Fig. 3. This flowchart systematically outlines the key steps, including model construction, physics setup, mesh generation, parameter configuration, and post-processing analysis.

TABLE I The size of the material used in different cable layers

Serial number	Main material physical parameters table		
	Layer name	*Materials*	*Thickness(mm)*
1	Piezoelectric patch	Lead zirconate titanate	1.0
2	Water buffer layer	Polyethylene	4.0
3	Insulating shield	Crosslinked polyethylene	1.0
4	Insulating layer	Crosslinked polyethylene	11.0
5	Conductor shield	Aluminum	1.25
6	Copper core	Copper	5.75

TABLE II Key parameters for materials used in different cable layers

Materials	Main material physical parameters table		
	Dielectric constant	*Density (g·cm⁻³)*	*Velocity of sound (m·s⁻¹)*
Aluminum	1000	2.7	6300
Copper	1.0	8.96	4700
Lead zirconate titanate	1.0	7.75	4000
Crosslinked polyethylene	2.3	0.92	2000
Polyethylene	2.35	0.93	2000

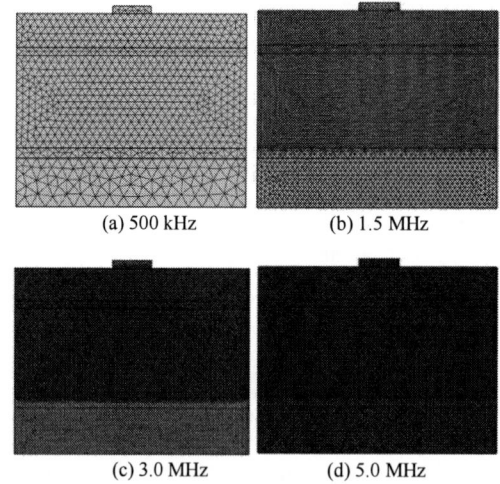

(a) 500 kHz (b) 1.5 MHz

(c) 3.0 MHz (d) 5.0 MHz

Figure 2. Meshing profiles for four different excitation frequencies

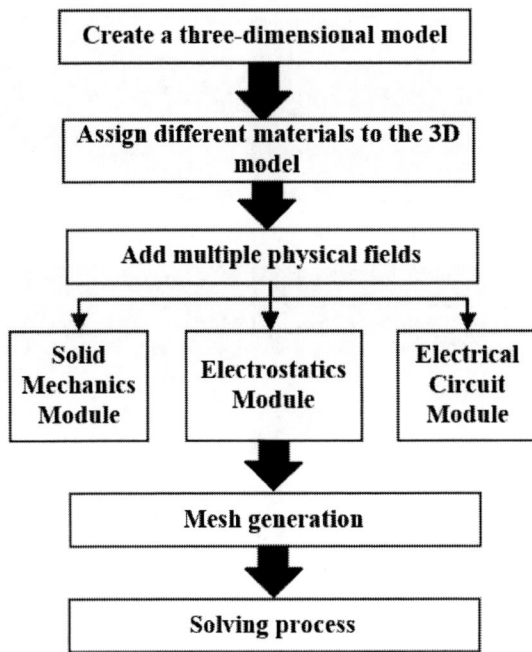

Figure 3. Ultrasonic Simulation Workflow Diagram

B. Multiphysics Coupling Calculations

This paper primarily investigates the coupling between solid mechanics and electrostatic fields, which together form the piezoelectric effect representing ultrasonic wave propagation. In this multi-physics coupling process, the interaction between solid mechanics and the electric field is crucial for the piezoelectric material's response to external excitation. The stress and strain in solid mechanics trigger mechanical responses through the material's elastic properties, whereas the changes in electric field strength and electric displacement in the electrostatic field affect the material's electrical response. They interact and transfer through the following coupling relationships.

$$T = c^E S - e^T E$$
$$D = eS + \varepsilon^S E \tag{1}$$

Where T represents the stress tensor, S is the strain tensor, E is the electric field strength, D is the electric displacement field, c^E is the elastic stiffness matrix under a constant electric field, e is the piezoelectric stress constant matrix, and ε^S is the dielectric constant matrix under constant strain [6].

Through the above coupling equations, mechanical stress not only affects the material's electric displacement response, but also the variation in the electric field in turn influences the strain state. This coupling effect forms the physical basis for piezoelectric materials to exhibit piezoelectric behavior under the combined action of external force and electric field [7, 8].

C. Boundary Conditions

This paper primarily studies the 2-D ultrasonic propagation simulations in high-voltage cables. Therefore, the electrostatic module is applied to the piezoelectric element, whereas the solid mechanics module is applied to the cable and the

piezoelectric element. The materials in each part of this module are all linear elastic materials.

The stress and strain in linear elastic material are proportional to each other. Linear materials have two distinct physical parameters, i.e., the elastic modulus (Young's modulus) E and Poisson's ratio v, which are the two physical parameters of linear materials.

For an isotropic linear elastic material, a set of three partial differential equations can be derived for the displacement vector u. The resulting Navier equation is expressed as follows.

$$(\lambda + \mu)\nabla(\nabla \cdot u) + \mu \nabla^2 u + f = \rho \frac{\partial^2 u}{\partial t^2} \tag{2}$$

Where λ and μ represent two independent material constants, referred to as the Lamé parameters.

Derived from E and v, Navier's equation can alternatively be expressed as following.

$$\frac{E}{2(1+v)}\left[\frac{1}{1-2v}\nabla(\nabla \cdot u) + \nabla^2 u\right] + f = \rho \frac{\partial^2 u}{\partial t^2} \tag{3}$$

In 2-D geometry, the stress in the material can be represented by the stress tensor, which takes the form of the following.

$$\sigma = \begin{bmatrix} \sigma_{xx} \sigma_{xy} \\ \sigma_{yx} \sigma_{yy} \end{bmatrix} \tag{4}$$

In the 2-D case, each component of the stress tensor corresponds to a force per unit area acting on the material. These components represent the force applied to the surface of the object, with one indicating the direction of the force and the other referring to the normal direction of the surface where the force is applied. From the perspective of torque equilibrium, this approach leads to a symmetric stress tensor with four independent components, namely σ_{xx}, σ_{xy}, σ_{yx} and σ_{yy}, where σ_{xy} is equal to σ_{yx}.

From the stress perspective, Newton's second law can be expressed as follows.

$$\rho = \frac{\partial^2 u}{\partial t^2} = \nabla \cdot S + Fv \tag{5}$$

Where ρ represents the material density, u is the initial displacement field value, t stands for time, S is the reference area, and F_v is the force component in the v-direction.

In the multi-physics coupling modeling of the piezoelectric effect, the boundary conditions of the electrostatic module play a crucial role in the accuracy of the simulation results. Therefore, it is essential to select the boundary conditions appropriately to accurately describe the physical phenomena in the piezoelectric effect. Common boundary conditions mainly include electrode boundaries, ground boundaries, and free boundaries. Each boundary condition has its specific physical meaning and mathematical expression [9, 10].

In this context, the free boundary condition refers to a situation where there is no externally applied voltage or current on the boundary, and the distribution of potential and electric field is determined by the internal and external mechanical and electrical interactions. For free boundaries, the following mathematical expression is typically used to describe the boundary condition:

$$\int D \cdot n dS = Q_0 \tag{6}$$
$$\frac{\partial Q_0}{\partial t} = I_{cir}$$

Where Q is the total charge on the electrode, D is the electric displacement field, n is the boundary normal vector, and I_{cir} is the current [11].

The grounding boundary is mainly described by the following mathematical expression.

$$V = 0 \tag{7}$$

Where V is the voltage applied at the boundary.

III. RESULTS AND DISCUSSIONS

To investigate the effects of different ultrasonic excitation frequencies on the propagation characteristics of high-voltage cables, this paper examined the propagation behavior of four excitation frequencies, i.e., 500 kHz, 1.5 MHz, 3.0 MHz, and 5.0 MHz, within hierarchical structure of high-voltage power cables. By comparing and analyzing the attenuation characteristics, propagation speed, and energy distribution of the ultrasonic waves at different frequencies, the paper reveals the impact of frequency on the propagation behavior of ultrasonic waves in high-voltage cables. The relevant research results are shown in Fig. 4 to Fig. 5.

As shown in Fig. 4, at a frequency of 500 kHz, the ultrasonic wave can clearly penetrate the conductor shielding layer and reach the copper core. However, as the frequency increases to 1.5 MHz, the ultrasound still propagates effectively to the copper core. When the frequency is further increased to 3.0 MHz, only a small portion of the ultrasound reaches the copper core. At 5.0 MHz, the ultrasonic wave can hardly penetrate the shielding layer to reach the copper core. This phenomenon indicates that as the frequency increases, the attenuation of the ultrasound within the conductor shielding layer significantly increases, leading to a substantial reduction in its propagation energy and a marked decline in its penetration capability.as shown in Fig. 5, at an excitation frequency of 500 kHz, the echo signal from the conductor shielding layer is weak and relatively unclear. As the frequency increases, the echo intensity gradually strengthens, and the clarity improves. However, at 1.5 MHz, the echo signal is difficult to distinguish from the noise. When the frequency increases to 3.0 MHz, the echo signal is significantly enhanced and becomes clearer. At 5.0 MHz, the enhancement of the echo signal tends to stabilize and reach a saturation point. This result indicates that the intensity and clarity of the echo signal improve with the increase in frequency, but at higher frequencies, it tends to reach a saturated level.

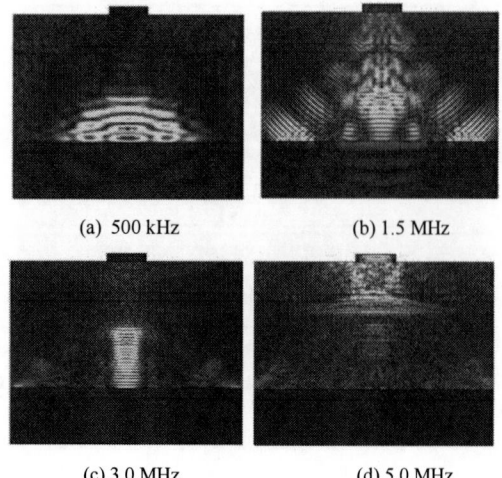

(a) 500 kHz (b) 1.5 MHz

(c) 3.0 MHz (d) 5.0 MHz

Figure 4. Propagation characteristics of ultrasound in the medium

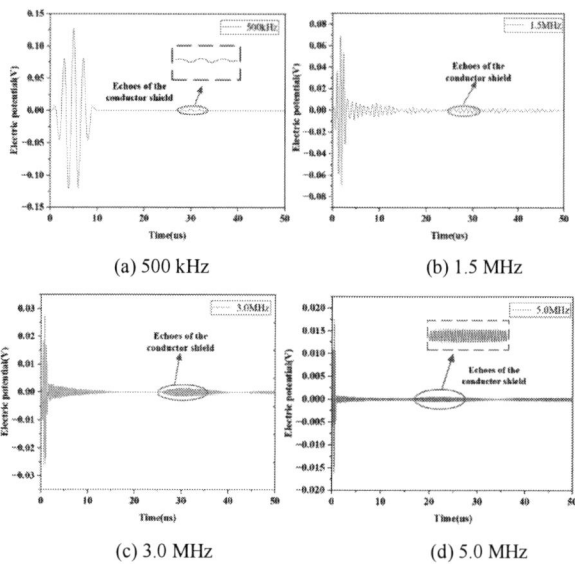

(a) 500 kHz (b) 1.5 MHz

(c) 3.0 MHz (d) 5.0 MHz

Figure 5. Ultrasonic echo distribution maps at four different frequencies

IV. CONCLUSIONS

Based on the COMSOL Multiphysics simulation platform, this paper establishes a finite element model of a high-voltage cable that includes the conductor shielding layer and copper core. The propagation characteristics and attenuation laws of ultrasound in multilayer media within the frequency range of 0.5-5.0 MHz are systematically studied through numerical simulation. The results show that within the frequency range of 0.5 MHz to 5.0 MHz, at 500 kHz, the ultrasonic wave has a longer wavelength and stronger penetration ability, allowing it to penetrate the conductor shielding layer and reach the copper core. However, as the frequency increases, the wavelength of the ultrasound gradually decreases, resulting in a reduced penetration capability and difficulty in effectively passing through the conductor shielding layer to reach the copper core. Additionally, at 500 kHz, the echo signal is weak and unclear, while at 3.0 MHz and 5.0 MHz, the echo signal intensity is significantly enhanced, and the clarity is improved. These results provide important theoretical insights for further research on ultrasonic defect detection in high-voltage cables, particularly regarding the relationship between the ultrasound's penetration characteristics and echo response at different frequencies. It should be noted that the finite element model established in this study still has certain limitations. The material parameters and geometric structures adopted are relatively idealized, without fully considering the influence of practical factors such as material defects and poor interfacial bonding under actual working conditions on ultrasonic wave propagation. Future research will incorporate more realistic material properties and service environments to further optimize the model and enhance its applicability and engineering reference value under complex conditions.

ACKNOWLEDGMENT

This work was supported by the 2023 Chongqing Municipal Educational Science Planning Project: *"Research and Practice on Digital Empowerment of the Curriculum System for Electrical Automation Technology Major"* (Project No. K23ZG3310207).

REFERENCES

[1] Peng X S, Wen J Y, Li Z H, et al. Rough set theory applied to pattern recognition of partial discharge in noise affected cable data[J]. IEEE Transactions on Dielectrics and Electrical Insulation, 2017, 24(1): 147-156.

[2] Yücel M K, Legg M, Kappatos V, et al. An ultrasonic guided wave approach for the inspection of overhead transmission line cables[J]. Applied Acoustics, 2017, 122: 23-34.

[3] Liu X, Wu B, Qin F, He C, Han Q, Configuration optimization of magnetostrictive transducers for longitudinal guided wave inspection in seven-wire steel strands. NDT & E Int 2010; 43(6):484-92.

[4] Li Fengzhong, Luo Ying. Damage Imaging of Lamb Wave in Isotropic Plate Using Phased Array Delay and Sum Based on Frequency-domain Inverse Scattering Model[J]. Nondestructive Testing and Evaluation, 2022, 37(6): 721-736.

[5] Lai Justine, Li Jing, Gniadecki Robert, et al. Gene Expression Profiling of Mycosis Fungoides in Early and Tumor Stage—A Proof-of-Concept Study Using Laser Capture/Single Cell Microdissection and NanoString Analysis[J]. Cells, 2021, 10(11): 3190-3190.

[6] Song Yongfeng et al. Enhanced Ultrasonic Flaw Detection Using an Ultrahigh Gain and Time-Dependent Threshold[J]. IEEE transactions on ultrasonics, ferroelectrics, and frequency control, 2018, 65(7): 1214-1225.

[7] Pankov A V, Tokar V L, Petronyuk Y S, et al.Determination of Fracture Toughness of Carbon Fiber Reinforced Plastics Free of the Crack Initiator Using Acoustic Microscopy[J].Inorganic Materials, 2022, 57(15): 1519-1524.

[8] Schafer ME, Spivak NM, Korb AS, Bystritsky A. Design, Development, and Operation of a Low-Intensity Focused Ultrasound Pulsation (LIFUP) System for Clinical Use. IEEE Trans Ultrason Ferroelectr Freq Control. 2021, 68(1): 54-64.

[9] Benammar Abdessalem, Drai Redouane, et al. Enhancement of Phased Array Ultrasonic Signal in Composite Materials using TMST Algorithm[J]. Physics Procedia, 2015, 70:488-491.

[10] Anand Chirag, Delrue Steven, et al. Simulation of Ultrasonic Beam Propagation from Phased Arrays in Anisotropic Media using Linearly Phased Multi Gaussian Beams[J]. IEEE transactions on ultrasonics, ferroelectrics, and frequency control, 2019, 67(1): 106-116.

[11] WangPrice Sharon S, Etibo Kristen N, Short Alicia P, et al. Validity and reliability of dry needle placement in the deep lumbar multifidus muscle using ultrasound imaging: an in-vivo study[J]. The Journal of manual & manipulative therapy, 2022, 30(5): 1-8.

2025 International Conference on Advanced Energy Systems and Power Electronics (AESPE 2025)

Parameter identification and SOC estimation of energy storage batteries based on HFA-AEKF

Longsheng Hua[1]

[1]Shenyang Ligong University, Shenyang, 110000, China

*Corresponding author's e-mail: 1853750970@qq.com

Abstract : **With the urgent need for high-precision energy storage management in intelligent manufacturing and mechatronic systems, this paper aims to improve the accuracy and robustness of lithium-ion battery parameter identification and State of Charge (SOC) estimation under dynamic operating conditions, to support the reliable operation of renewable energy integration and intelligent equipment. Existing methods for dynamic parameter identification suffer from issues such as rigid historical data weighting and inadequate global search capabilities. To address these challenges, this paper proposes a Hybrid Firefly Algorithm (HFA), which combines the dynamic weight adjustment capability of Adaptive Forgetting Factor Recursive Least Squares (AFFRLS) with the global optimization capability of the Firefly Algorithm (FA), significantly enhancing the parameter identification accuracy of the second-order RC model. Furthermore, by integrating Adaptive Noise Extended Kalman Filtering (AEKF), a joint estimation framework of HFA-AEKF is established. Simulation experiments demonstrate that the new method, through dynamically balancing the contribution of new and old data and multi-dimensional parameter collaborative optimization, reduces voltage fitting errors by 16% and 31.5%, and decreases the Root Mean Square Error (RMSE) of SOC estimation by 16.7% and 32.4%, compared to FA-AEKF and AFFRLS-AEKF, respectively. This shows superior adaptability to operating conditions and nonlinear processing capabilities. The findings of this study provide a new method for precise energy management of intelligent manufacturing equipment, effectively extending the lifespan of energy storage systems and enhancing the dynamic response reliability of mechatronic devices, contributing to the deep integration of renewable energy and intelligent factories.**

Keywords—Energy Storage Battery; State of Charge (SOC); Hybrid Firefly Algorithm (HFA); Adaptive Extended Kalman Filter (AEKF); Parameter Identification; Second-Order RC Model

I. INTRODUCTION

High-precision energy management is crucial for ensuring the reliable operation of high-end equipment, such as industrial robots and intelligent production lines. The accuracy of lithium-ion battery state of charge (SOC) estimation directly impacts energy scheduling and dynamic response capabilities of these devices.

Existing lithium battery parameter identification methods can generally be classified into three types: direct measurement, offline identification, and online identification [1]. Among these, online recursive algorithms have gained significant attention due to their real-time advantages, such as the improved Kalman filter framework, which implements recursive optimal estimation via a state-space model [2]. Literature [3] employs a dual-extended Kalman algorithm to jointly optimize SOC estimation and parameter calibration, demonstrating the

effectiveness of this approach. In literature [4], the AFFRLS-AEKF method performs well under steady-state conditions, but in intermittent charging and discharging scenarios in intelligent factories, the voltage fitting error accumulates to 17.2%. Literature [5] shows that the FA-EKF method suffers from time-varying noise mismatches, leading to an RMSE over 3% for SOC estimation, which fails to meet the high-precision requirements for mechatronic servo control. This paper proposes an energy storage battery parameter-state synergistic optimization framework incorporating the Hybrid Firefly Algorithm (HFA). By integrating the dynamic forgetting factor adjustment of AFFRLS with the global search capabilities of FA, the proposed method enables real-time updates of the parameter search space and dynamic balancing of data weights. This solution overcomes the parameter lag issue caused by sudden changes in operating conditions and establishes a closed-loop HFA-AEKF optimization system, breaking through the dynamic coupling modeling limitations of traditional decoupled methods. Simulation results show that this method significantly reduces voltage fitting errors compared to FA-AEKF and AFFRLS-AEKF, with more stable SOC estimation, providing a high-precision battery model for digital twin-driven intelligent production lines.

This research achieves a dual breakthrough in the fields of intelligent manufacturing and mechatronics. Technologically, the HFA-AEKF framework improves dynamic identification accuracy and estimation stability, setting a new standard for real-time energy management in AGVs, industrial robots, and other devices. Systematically, the closed-loop optimization enhances the adaptability of energy storage systems to intelligent manufacturing conditions and reduces equipment downtime risks. Experimental results demonstrate that this method extends lithium battery lifespan, reduces response delays, provides technical support for the integration of renewable energy and intelligent manufacturing, enhances industrial system energy efficiency, and contributes to achieving the "carbon neutrality" goal.

II. LITHIUM-ION BATTERY MODELING AND PARAMETER IDENTIFICATION

A. Lithium-ion battery equivalent circuit model

First, Electronic elements like impedances and condensers can effectively replicate the external voltage characteristics of a battery, and their parameters can be identified through experimental methods. Considering the trade-off between considering model fidelity and algorithmic intricacy, the second-order RC surrogate circuit framework is opted for. for SOC estimation in this study, as illustrated in Figure 1:

979-8-3315-2490-6/25 $31.00 © 2025 IEEE

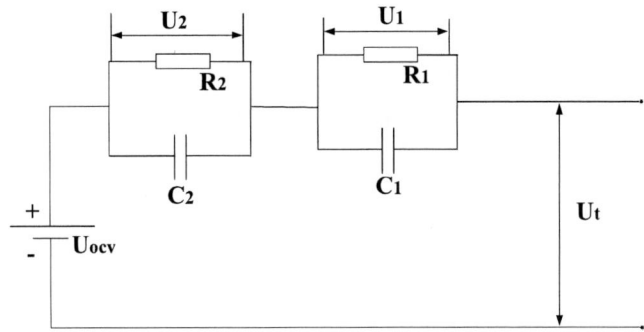

Figure 1. Second-Order RC Equivalent Circuit Model.

In Figure 1, U_{ocv} and U_t denote the no-load potential and electrode potential of the battery., respectively. R_0 represents the resistive internal impedance; R_1 XX is the electrochemical component polarization resistance; R_2 is the concentration polarization resistance; C_1 is the electrochemical polarization capacitance; C_2 is the concentration polarization capacitance; U_1 and U_2 are the electrochemical polarization voltage and concentration polarization voltage, respectively, and I(t) is the current at time [6].

At the same time, this paper also considers the impact of battery temperature and SOC on the model parameters. The RC structure is used to simulate the battery voltage relaxation process, where the polarization processes of the battery are distinguished. Based on the equivalent circuit model in Figure 1, the discrete state equations and SOC recursive calculation formulas for the battery system are derived.

$$U_1(k) = \exp\left(\frac{-T_s}{R_1 \times C_1}\right) \times U_1(k-1) + R_1 \times \left[1 - \exp\left(\frac{-T_s}{R_1 \times C_1}\right)\right] \times I(k-1)$$
(1)

$$U_2(k) = \exp\left(\frac{-T_s}{R_2 \times C_2}\right) \times U_2(k-1) + R_2 \times \left[1 - \exp\left(\frac{-T_s}{R_2 \times C_2}\right)\right] \times I(k-1)$$
(2)

$$U_t(k) = U_{OCV}(k) - U_1(k) - U_2(k) - R_0 \times I(k)$$

$$S_{OC}(k) = S_{OC}(k-1) - \frac{T_s \times I(k)}{C_a} \times 100\%$$
(3)

B. Model parameter identification based on hybrid firefly optimization algorithm (HFA)

For the dynamic parameter identification problem of the second-order RC surrogate circuit framework., the HFA integrates AFFRLS and the FA to construct a collaborative optimization framework. Its core functions are as follows:

(1) AFFRLS Module: Dynamic Data Weight Balancing

The AFFRLS module introduces the Sliding Window Coefficient of Variation (SWCV) to dynamically evaluate parameter fluctuations in real time. By adjusting the forgetting factor, it balances the weight between new and historical data[7]:

$$\lambda_k = \begin{cases} \lambda_{k-1} - \Delta\lambda, & \text{if } CV_k < CV_{low} \\ \lambda_{k-1} + \Delta\lambda, & \text{if } CV_k > CV_{high} \\ \lambda_{k-1}, & \text{otherwise} \end{cases}$$
(4)

$CV_k = \frac{\sigma_{\bar{\mu}}}{\mu_{\bar{\mu}}}$ represents the coefficient of variation of the parameter estimate. $N = 10$ denotes the sliding window size.

(2) FA Optimization Module: Swarm Intelligence Global Search

Based on the firefly brightness attraction mechanism, the attraction of firefly i to firefly j is defined as:

$$\beta_{ij} = \beta_0 \cdot e^{-\gamma r_{ij}^2} \cdot \frac{I_j - I_i}{|I_j - I_i| + \epsilon}$$
(5)

Where $r_{i,j}$ is the Euclidean distance between fireflies. $\gamma = 0.001$ is the light absorption coefficient. $\epsilon = 10^{-6}$ prevents the denominator from being zero. A dynamic step size factor $\alpha_t = \alpha_0 \cdot (0.97)^t$ and a perturbation term $\tilde{\epsilon}_i \sim U(-0.5, 0.5)$ are incorporated to augment the universal exploration capacity. and avoid local optima.

(3) Collaboration Mechanism: Parameter-Search Space Closed-Loop Feedback

Initial Population Generation The real-time parameter estimates output from AFFRLS are used as the center of the FA initial population, generating neighboring solutions:

$$x_i^{(0)} = \hat{\theta}_{AFFRLS} + \delta \cdot \text{rand}(-1,1)$$
(6)

Where the neighborhood radius $\delta = 0.2 \cdot \left\| \hat{\theta}_{AFFRLS} \right\|$ ensures that the search range is compatible with the scale of the parameters.

Objective Function Design: The optimization objective is the voltage mean squared error (VMSE), combined with a dynamic weight coefficient:

979-8-3315-2490-6/25 $31.00 © 2025 IEEE

$$J = \sum_{k=1}^{N} w_k \cdot \left(U_t(k) - \hat{U}_t(k) \right)^2 \tag{7}$$

Where $w_k = \dfrac{1}{\sigma_U^2(k)}$ is dynamically adjusted based on the voltage measurement noise variance, suppressing error accumulation in high-noise regions.

III. AEKF ALGORITHM FOR BATTERY SOC ESTIMATION

By employing this approach, EKF can update the noise covariance matrices in real-time during execution, effectively handling uncertainties and external disturbances in the actual system, thereby improving the accuracy and stability of SOC estimation. The specific computational process of the AEKF algorithm is as follows[8]:

(1) Initialization: Setting Initial Values

(2) Prior Estimation: Perform system state and error covariance prediction;

$$\hat{X}_k^- = f(X_{k-1}, u_{k-1}), \quad P_k^- = A_{k-1} P_{k-1} A_{k-1}^T + Q_{k-1} \tag{8}$$

(3) Posterior Estimation: Perform state updates and correct the system measurement values to obtain the state estimate and covariance estimate results, while updating the Kalman gain and noise covariance. The algorithm defines the innovation at time k as the discrepancy between the true value and the observed value:

$$e_k = y_k - C_k \hat{X}_k^- \tag{9}$$

Update the Kalman Gain Matrix:

$$L_k = P_k^- C_k^T (C_k P_k^- C_k^T + R_k)^{-1} \tag{10}$$

The innovation sequence Incorporates not only the instantaneous system disturbance data but also the real-time observation noise data. Based on this new information, adaptive noise covariance matching can be performed in real-time. Correct the system state and error covariance:

$$\hat{X}_k^+ = \hat{X}_k^- + L_k e_k, \quad P_k^+ = (E - L_k C_k) P_k^- \tag{11}$$

IV. SIMULATION VALIDATION

The accuracy of parameter identification is crucial for precise SOC estimation. This paper introduces the parameters identified by the optimization algorithm into the RC equivalent circuit model. By comparing the estimated terminal voltage with the actual terminal voltage, the tracking performance of the model parameters identified by HFA was validated. Figure 2(a), (b), and (c) show the comparison results of the estimated terminal voltage and actual terminal voltage for AFFRLS, FA, and HFA algorithms respectively, where the actual voltage was obtained using the open-circuit method.

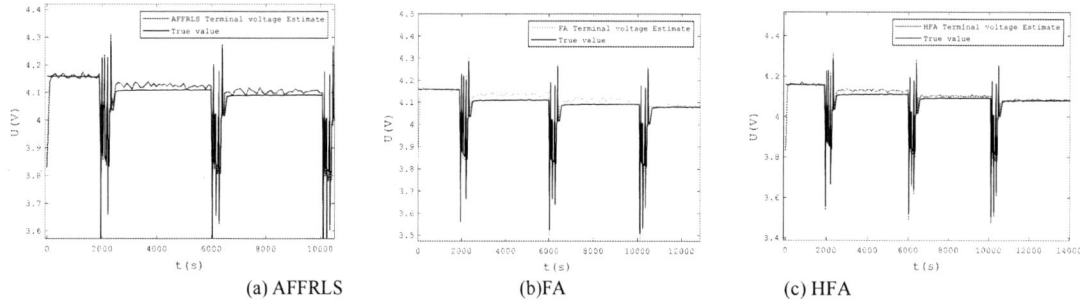

| | (a) AFFRLS | (b)FA | (c) HFA |

Figure 2. Comparison of estimated and actual voltage values for AFFRLS, FA, and HFA algorithms.

It can be seen from Figure 2(a), (b), and (c) that the Hybrid Firefly Optimization Algorithm (HFA) provides more accurate voltage estimation compared to FA and AFFRLS methods. Additionally, the error of the HFA algorithm is generally more stable, while the errors of the other two algorithms exhibit larger fluctuations. The results demonstrate that HFA outperforms AFFRLS and FA in voltage tracking.

Table 1 shows the average absolute error, root mean square error, and average absolute error percentage between the estimated terminal voltage and the actual terminal voltage. The Hybrid Firefly Algorithm reduces the root mean square error by 16% and 31.5% compared to the other two algorithms, respectively.

Table 1 Comparison of voltage errors between AFFRLS, FA, and HFA algorithms.

Error Parameters	AFFRLS	FA	HFA
MAE	7.1×10^{-3}	6.5×10^{-3}	4.2×10^{-3}
RMSE	9.2×10^{-3}	7.5×10^{-3}	6.3×10^{-3}
Percentage of mean absolute error (%MAE)	1.7×10^{-3}	1.6×10^{-3}	1.5×10^{-3}

Grounded on the parameter discernment outcomes of the second-order RC surrogate circuit framework. The data used for the simulation experiment was the voltage and current measured during the composite pulse power experiment of the pulse testing. This paper uses the voltage and current data of the lithium battery model 18650 under 0°C discharge conditions. The saturation voltage is 4.8V, the cutoff voltage is 2V, and the discharge current is 0.5A. The main parameters are shown in

Table 2.

Table 2 Unloaded potential and charge state parameters of lithium cells.

SOC	Uoc/V
0	1.979
0.1	2.167
0.2	2.651
0.3	3.159
0.4	3.859
0.5	4.623

The coefficients of the determined second-order RC surrogate circuit framework are incorporated into the AEKF simulation model. The AEKF is then used to estimate the SOC of the battery, as shown in Figure 3. The figure compares three state estimation methods: FA-AEKF, AFFRLS-AEKF, and HFA-AEKF. From the comparison, it can be seen that the SOC estimation results of the HFA-AEKF algorithm are closer to the true values.

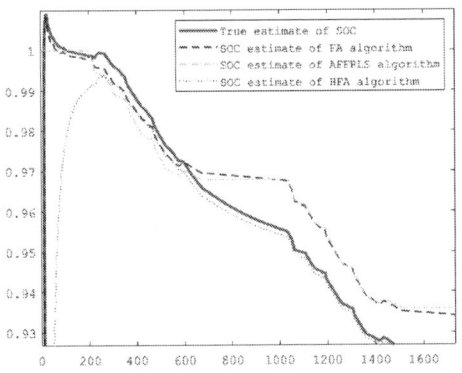
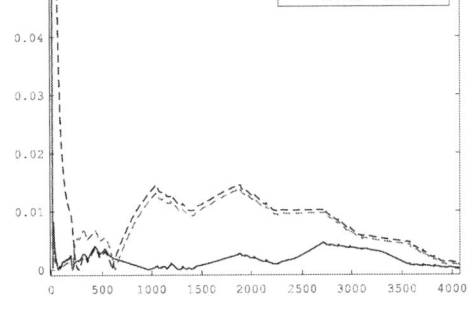

(a)Comparison of SOC errors for three algorithms (b)Comparison of three algorithms SOC estimates

Figure 3.   Comparison of SOC estimates and SOC estimation errors for three algorithms

As shown in Figure 3(a), the comparison of SOC estimation performance among the FA-AEKF, AFFRLS-AEKF, and HFA-AEKF algorithms reveals that the SOC estimation results of the HFA-AEKF algorithm are closer to the true values.

Figure 3(b) further presents the SOC estimation errors of the FA-AEKF, AFFRLS-AEKF, and FA-AEKF algorithms. The results show that the SOC estimation error of the HFA-AEKF algorithm is smaller and its fluctuations are more stable.

Table 3. Comparison of SOC estimations using AFFRLS-AEKF, FA-EKF, and HFA-EKF.

Error Parameters(SOC)	AFFRLS-AEKF	FA-EKF	HFA-EKF
MAE(SOC)	$7.8×10^{-3}$	$6.5×10^{-3}$	$6×10^{-3}$
RMSE(SOC)	$8.4×10^{-3}$	$7.9×10^{-3}$	$7×10^{-3}$
Percentage of Mean Absolute Error (%MAE) (SOC)	$5×10^{-2}$	$4.2×10^{-2}$	$3.8×10^{-2}$

As shown in Table 3, the SOC estimation result of the AFFRLS-AEKF algorithm has an average absolute error of $7.8×10^{-3}$, the SOC estimation result of the FA-AEKF algorithm has an average absolute error of $6.5×10^{-3}$, and the SOC estimation result of the HFA-AEKF algorithm has an average absolute error of $6×10^{-3}$. The errors were reduced by 17% and 23%, respectively. The root mean square error decreased by 6% and 16.7%, and the mean absolute error percentage decreased by 16% and 24%. The HFA-AEKF algorithm demonstrates a more significant advantage.

V. CONCLUSION

This study addresses the challenges of dynamic parameter identification lag and state estimation errors in lithium batteries under intelligent manufacturing scenarios. A collaborative optimization framework based on the Hybrid Firefly Algorithm (HFA) and Adaptive Extended Kalman Filtering (AEKF) is proposed, aiming to improve the accuracy of energy storage battery models and the robustness of State of Charge (SOC) estimation, providing technical support for efficient energy management in intelligent equipment. Simulation results show that the HFA algorithm, by integrating dynamic forgetting factor adjustment with a global search mechanism, significantly improves parameter identification accuracy. The AEKF estimator based on parameters identified by HFA exhibits stronger stability under varying load conditions, and its SOC estimation accuracy is notably superior to traditional methods, demonstrating significant advantages in extending battery life

979-8-3315-2490-6/25 $31.00 © 2025 IEEE

and improving dynamic response performance. This research provides a new solution for intelligent manufacturing and mechatronics, enhancing the reliability of energy management for AGVs, industrial robots, and other equipment under complex conditions. The limitation of this study lies in its failure to address extreme environments and battery aging. Future work will focus on hardware-in-the-loop validation, multi-physics modeling, and lightweight algorithm design, promoting the application of this method in digital twin factories, and supporting intelligent equipment energy efficiency optimization and the achievement of carbon neutrality goals.

REFERENCES

[1] WEI J, CHEN C, DONG G. Global Sensitivity Analysis for Impedance Spectrum Identification of Lithiuml on Batteries Using Time-Domain Response[J]. IEEE Transactions on IndustrialElectronics,2022,70(4):3825-3835.

[2] WANG H Y, DUAN J, H, Review of Lithium-Thionyl Chloride Battery SOC Estimation Methods [J]. Instrumentation Technology.

[3] JIANG C, WANG S, WU B, et al. A state-of charge estimation method of the power lithium-ion battery in complex conditions based on adaptive square root extended Kalman filter [J].Energy2021,219:119603

[4] GE K, Battery SOC Estimation Method Based on AFFRLS-AEKF [J]. Information Technology and Informatization. 2024, (12):179-185.

[5] LI X C, YANG F, NA C N. Parameter Identification and SOC Estimation of Lithium Batteries Based on FA-EKF [J]. Electrical Engineering and Electronics. 2024,(12):9-14.

[6] XING S, LI G, ZHANG Y J, et al. Online diagnosis of state of health for lithium-ion batteries based on short-term charging profiles[J]. Journal of power sources, 2020, 471(9)228478.

[7] WANG S, TAKYI-ANINAKWA P, JIN S, et al. An improved feedforward-long short-term memory modeling method for the whole-life-cycle state of charge prediction of lithium-ion batteries considering current-voltage-temperaturevariation1]. Energy, 2022, 254(9):124224.

[8] GUO Y, ZHAO Z, HUANG L. SoC estimation of lithium battery based on AEKF algorithm[]. Energy Procedia , 2017105(5):4146-4152.

AI Optimization for Coordinated Scheduling of Distributed Energy Storage and Renewable Energy at Multiple Time Scales

Junkai Wu
Xiamen University of Technology
Xiamen, China
junkaiwu2020@163.com

Abstract—**With the continuously increasing penetration rate of renewable energy in power systems, how to achieve efficient collaborative dispatching between distributed energy storage and renewable energy has become an urgent problem to solve. This paper focuses on the collaborative dispatching of distributed energy storage and renewable energy under multiple time scales, aiming to enhance the economic efficiency and stability of the system through artificial intelligence (AI)-based optimization algorithms. This research constructs a day-ahead/intraday two-layer dispatch framework to address long-term forecasting and short-term adjustment requirements. An AI-driven prediction and allocation system is developed, where forecasting modules accurately project renewable generation and load demand while optimization modules achieve resource allocation through intelligent algorithms. Experimental results demonstrate substantial dispatch cost reduction and enhanced system flexibility, facilitating large-scale integration of renewables and distributed storage.**

Keywords—Multi-time scale; Distributed energy storage; New energy; Collaborative dispatching; AI optimization algorithm

I. INTRODUCTION

Rising global energy demand and escalating environmental priorities have positioned distributed energy storage and renewable technologies as critical power system components. The geographic adaptability of distributed storage mitigates renewable intermittency, enhancing grid stability and economic performance [1]. Concurrently, massive renewable integration presents unprecedented regulatory challenges — manifesting through heightened load volatility and uncertainty — while demanding efficient storage utilization for system-wide optimization [2].

To be specific，multi-time-scale scheduling emerges as a foundational solution. Coordinating distributed storage with renewables across day-ahead to intraday horizons enables superior operational harmony. During day-ahead optimization, meteorological forecasts (wind speed/solar irradiance) preemptively shape storage charge/discharge planning — curtailing operating expenses and active power losses [1]. Intraday optimization then refines storage operations to track real-time generation-load fluctuations. This hierarchical strategy elevates system responsiveness and adaptability [3].

The implementation of AI optimization algorithms substantially augments coordinated scheduling performance. Deep learning and reinforcement learning techniques deliver enhanced accuracy in predicting renewable energy outputs and electricity consumption patterns, delivering critical input signals for distributed storage charge/discharge dispatch strategies [4]. Notably, reference [5] introduced an online battery health monitoring approach integrating multi-time-scale aging dynamics. Research outcomes confirm that deep learning frameworks particularly elevate battery state estimation precision. Correspondingly, studies [6] and [7] respectively employed convolutional neural networks alongside adaptive multi-scale joint estimation methodologies to accomplish refined battery state modeling. These investigations collectively demonstrate AI technology's significant potential within renewable-integrated distributed storage coordination systems.

Addressing renewable energy's inherent unpredictability, this research develops an artificial intelligence forecasting approach leveraging multi-temporal feature analysis. The method deciphers dynamic patterns in energy storage status and consumption needs, delivering high-fidelity forecasts for coordinated dispatch. A deep learning architecture additionally refines modeling of distributed storage degradation dynamics and partial cycling behavior. Experimental validation confirms the framework's operational versatility across scenarios alongside measurable system performance gains.

II. MULTI-TIME-SCALE OPTIMIZATION SCHEDULING MODEL

A. Short-term Optimization Scheduling Model

The short-term scheduling model serves as a cornerstone for multi-timescale renewable-storage coordination. Targeting operational cost reduction and active power loss minimization, it establishes efficient power system management. Our dual-objective optimization model reconciles economic and technical factors, incorporating distinctive renewable and storage attributes.

Distributed storage systems offer crucial short-term scheduling flexibility through rapid response capabilities. Modeling necessitates examining renewable characteristics — particularly wind and solar generation. Wind power modeling typically employs a two-parameter Weibull distribution correlated with wind velocity [8], as follows:

$$P_w(V) = \frac{k}{c}\left(\frac{V}{c}\right)^{k-1} \exp\left[-\left(\frac{V}{c}\right)^k\right] \quad (1)$$

where V is the wind speed at the wind turbine; c and k are the distribution parameters and shape parameters. The output power of photovoltaic systems is significantly affected by environmental factors such as light intensity and temperature.

For effective coordinated scheduling, models must account for storage cycling behaviors and renewable intermittency. Storage units charge during off-peak tariff intervals and discharge during peak pricing, lowering operational expenditures and stabilizing load variations. Integration of AI optimization enhances renewable generation forecasting precision, consequently refining scheduling decisions. Deep learning-driven multi-timescale feature extraction applied to battery health assessment could help boost system resilience in complex environments.

In terms of optimization objectives, we set the minimum operation cost and the minimum active power network loss as the dual constraints, and design the corresponding mathematical expressions. The operation cost function can be expressed as:

$$C = \sum_{t=1}^{T} [C_g(P_g(t)) + C_s(P_s(t))] \qquad (2)$$

where C_g and C_s represent the generation cost of traditional units and the scheduling cost of the energy storage system; $P_g(t)$ and $P_s(t)$ represent the power output of traditional units and the energy storage system in each period.

Comprehensive investigation of AI-enhanced short-term scheduling achieves dual techno-economic optimization. This methodology tackles renewable integration uncertainties while supporting subsequent intraday scheduling refinement.

B. Day-ahead Optimization Scheduling Model

As a strategic component of multi-timescale scheduling, the day-ahead model corrects renewable generation and load forecast deviations. This approach lowers operating costs and strengthens system reliability.

Coordinating distributed storage with renewables proves essential here. DES flexibility enables multi-timescale regulation against renewable intermittency. Through rapid energy absorption/release, DES smoothens instantaneous wind/PV fluctuations. Meanwhile, AI algorithms substantially improve scheduling efficacy—accurately projecting renewable outputs and loads to optimize storage cycling strategies.

Specifically, the objective function of the real-time optimal dispatch model is usually set to minimize the correction cost, while taking into account the safety, stability and economy of the system. The correction cost covers the additional costs caused by the deviation of the output of renewable energy and the loss costs of the operation of energy storage. To achieve this goal, constraints are introduced in the model to ensure the physical feasibility of the system. The charging and discharging power of the energy storage system must satisfy the following constraints:

$$P_{charge}^{min} \leq P_{charge}(t) \leq P_{charge}^{max} \qquad (3)$$

$$P_{discharge}^{min} \leq P_{discharge}(t) \leq P_{discharge}^{max} \qquad (4)$$

Meanwhile, the state of charge (SOC) of distributed energy storage also needs to fluctuate within a reasonable range to avoid overcharging or over-discharging phenomena [3][6]. This can be described by the SOC dynamic equation:

$$SOC(t+1) = SOC(t) + \eta_{charge} \cdot P_{charge}(t) \cdot \Delta t - \frac{1}{\eta_{discharge}} \cdot P_{discharge}(t) \cdot \Delta t \qquad (5)$$

where η_{charge} and $\eta_{discharge}$ represent the charging and discharging efficiencies respectively, and Δt represents the time step.

For the AI optimization algorithms in intraday scheduling, advanced technologies such as deep learning can achieve accurate prediction of the output of renewable energy and combine the regulation capacity of energy storage to formulate the optimal scheduling strategy. Capsule neural networks based on the attention mechanism of multiple time scales can capture the dynamic characteristics at different time scales, thereby enhancing the robustness and adaptability [4]. Model predictive control methods are also widely used in the optimization scheduling of distributed energy storage to achieve the organic combination of real-time regulation and long-term planning [3].

It is worth noting that the intraday optimization scheduling model also needs to consider the actual operating environment of the power grid, such as voltage stability, line power flow restrictions, etc. This makes the model design more complex, but also better adapts to the actual application requirements. Embedding cooperative storage-renewable scheduling within AI optimization frameworks allows intraday models to meaningfully cut operating expenses and upgrade overall grid performance [8] [9].

III. AI OPTIMIZATION ALGORITHM DESIGN

A. AI Prediction Algorithm

In multi-timescale coordination challenges, AI forecasting constitutes fundamental scheduling infrastructure. It handles renewable unpredictability (wind/solar outputs) and load variation impacts on grid stability. This segment explores the prediction algorithm's development process.

The algorithm identifies historical/real-time data patterns to project future renewable yields, load requirements, and storage conditions. Based on these predictions, the optimization scheduling algorithm can formulate more reasonable energy storage charging and discharging strategies, thereby improving energy utilization efficiency and reducing operating costs.

1) Data-driven prediction models

AI prediction algorithms are typically constructed based on data-driven methods. Common models include machine learning algorithms and deep learning algorithms. Specifically, time series prediction methods or regression analysis models (such as random forests, support vector machines) can be adopted. Taking Long Short-Term Memory Network (LSTM) as an example, this model has the ability to handle time series data and is highly suitable for predicting renewable energy generation power and user load demand.

In practical applications, AI prediction algorithms need to be trained and validated with multi-source data. For the power generation prediction of photovoltaic power stations, the following data can be relied on:

- Meteorological data: covering solar irradiance, temperature, humidity, etc., environmental parameters.

979-8-3315-2490-6/25 $31.00 © 2025 IEEE

- Historical power generation data: long-term records of photovoltaic power station power generation, capturing seasonal and diurnal periodic changes.

- Electricity load data: the electricity demand patterns of users also affect the operation status of photovoltaic systems.

2) Prediction process design

The implementation of the AI prediction algorithm typically involves the following steps:

- Data preprocessing: Data cleaning is the first step of AI prediction, aiming to remove noise and outliers, and fill in missing data. In photovoltaic power generation prediction, if the irradiance data for a certain time period is missing, interpolation methods can be used for filling.

- Feature engineering: Transforming and extracting the original data to generate feature variables that are helpful for prediction. For the power generation prediction of wind farms, key features such as wind speed and wind direction can be extracted.

- Model training and validation: Selecting appropriate AI models for training and cross-validating their performance. Training data usually comes from actual operation records over the past few years, while validation data is selected from recent datasets.

- Output and correction of prediction results: Based on the trained model, output the prediction results, and combine actual conditions to correct the prediction values. In load prediction, a real-time feedback mechanism can be used to dynamically adjust the prediction errors.

3) Case analysis of actual applications

Taking a regional power grid as an example, this region contains multiple distributed photovoltaic power stations and wind farms, with a total installed capacity of renewable energy generation of approximately 100MW. To achieve precise prediction, researchers used a deep learning model based on LSTM, with training data covering the past three years of meteorological and power generation records. After multiple iterations and optimizations, the prediction accuracy of the final model reached over 93%, with the hourly power generation prediction error of photovoltaic power stations controlled within ±6%.

At the same time, load prediction is also one of the important application areas of AI prediction algorithms. In this regional power grid, the research team developed a hybrid model combining XGBoost and LSTM for predicting future 24-hour user load demand. The results showed that the Mean Absolute Percentage Error (MAPE) of the hybrid model was only 4.1%, significantly outperforming the prediction effect of a single model.

TABLE I. CASE ANALYSIS

Data Type	Numeric Value	Remarks
Total installed capacity of renewable energy power	100 MW	Distributed photovoltaic power stations and wind farms
Prediction accuracy of LSTM model	93%	Trained based on past three years' meteorological and power generation records
Hourly power generation prediction error	±6%	Prediction results of LSTM model
MAPE	4.1%	Prediction results of XGBoost and LSTM hybrid model

B. AI Scheduling Algorithm

In the context of multiple time scales, the AI scheduling algorithm is one of the core technologies for achieving coordinated scheduling of distributed energy storage and renewable energy. This section focuses on discussing the principles, design and performance of the AI scheduling algorithm in practical applications. The AI scheduling algorithm integrates multi-time scale data and takes into account the charging and discharging characteristics of distributed energy storage systems and the uncertainty of renewable energy generation, providing an intelligent solution.

The AI scheduling algorithm needs to have the ability to handle multi-time scale data. Literature [3] proposed a multi-time scale control strategy based on model predictive control, which has been successfully applied in the hybrid microgrid to smooth output power fluctuations. Similarly, in the coordinated scheduling of distributed energy storage and renewable energy, the AI scheduling algorithm needs to manage both day-ahead and intraday scheduling tasks simultaneously. Day-ahead scheduling focuses on global optimization and usually aims to reduce operating costs and reduce active network losses [1]; while intraday scheduling focuses on minimizing short-term correction costs and ensuring the real-time and accuracy of scheduling instructions [5]. Combining these two aspects forms a complete scheduling framework.

Secondly, the AI scheduling algorithm needs to address the small capacity and high flexibility characteristics of distributed energy storage systems. Literature [10] studied the technical feasibility of distributed energy storage devices participating in primary frequency modulation of the power grid, indicating that its fast response speed can provide millisecond-level power regulation capabilities for the scheduling system. Combining these characteristics, the AI scheduling algorithm can aggregate multiple distributed energy storage units to form a virtual large-scale energy storage system, thereby enhancing the overall control capability. To achieve more refined control, the algorithm needs to support dynamic adjustment of the working status of each energy storage unit to adapt to different load demands and changes in renewable energy generation conditions [8].

Thirdly, the AI scheduling algorithm also needs to consider the uncertainty of renewable energy generation. The randomness and intermittency of wind power and photovoltaic power generation pose challenges to scheduling. To this end, some advanced AI algorithms have been introduced, such as deep learning and reinforcement learning methods. Literature [4] proposed a capsule neural network that integrates a multi-scale feature attention mechanism. This algorithm can extract text features from different levels and, similarly, can be applied to renewable energy generation prediction to capture its complex spatiotemporal distribution patterns [6].

The actual effectiveness of the AI scheduling algorithm depends on high-quality data support and efficient computing platforms. Literature [3] emphasized the importance of environmental space control, which is also applicable to the energy scheduling field. A complete AI scheduling system should cover data collection, processing and analysis, as well as the final execution control links [7]. The integration of cloud computing technology and edge computing provides new possibilities for large-scale distributed energy storage and renewable energy coordinated scheduling [2].

Consequently, as a principal multi-timescale scheduling instrument, the AI algorithm should cover day-ahead/intraday objectives while addressing distributed storage operational traits and renewable uncertainties.

IV. RESULT ANALYSIS AND DISCUSSION

Validation experiments assess multi-timescale AI scheduling performance using actual grid datasets. Tests include day-ahead and intraday phases, demonstrating time-scale adaptability.

Day-ahead testing uses historical data, load projections, and meteorological forecasts to predict renewables and optimize storage cycling. The 24-hour hourly evaluation compares algorithmic outputs with operational records for accuracy and cost-efficiency assessment.

Intraday testing applies rolling optimization with 15-minute adjustments. It introduces stochastic disturbances — say unexpected load spikes or renewable fluctuations—to mimic real-grid uncertainty factors. The results of the intraday optimization scheduling are analyzed to evaluate the robustness and response speed of the algorithm.

The experiment also sets up multiple comparison scenarios, covering scheduling schemes without energy storage participation and AI prediction (case 1), single energy storage scheduling schemes with no AI prediction (case 2), and distributed energy storage aggregation scheduling schemes with AI prediction (case 3). The comparative analysis of these scenarios further verifies the advantages of the proposed coordinated scheduling.

TABLE II. PARAMETER SETTING

Data category	Parameter	Value or description
Load data	Time span	3 months (summer peak period)
Load data	Time resolution	15 minutes
Load data	Maximum load	100MW
Load data	Minimum load	30MW
Wind farm data	Total installed capacity	50MW
Wind farm data	Average wind speed	7m/s
Wind farm data	Standard deviation	2m/s
Photovoltaic power station data	Total installed capacity	30MW

Distributed energy storage data	Total energy storage capacity	20MWh (consisting of 20 distributed energy storage units, each with 1MWh)
Distributed energy storage data	Charging and discharging efficiency	90%
Distributed energy storage data	Maximum charging and discharging power	10MW
Weather data	Time resolution	1 hour
Operating cost data	Cost of starting and stopping thermal power units	Approximately 300 yuan/MW per start-stop
Operating cost data	Cost of energy storage charging	0.4 yuan/kWh
Operating cost data	Cost of energy storage discharging	0.6 yuan/kWh

TABLE III. NUMBER RESULTS

Scenario	Indicator	Improvement Effect
Stabilize load fluctuations	Reduce load fluctuation amplitude	About 28%
Enhance the capacity of absorbing renewable energy sources	Reduce the rate of discarding light energy	From 16% to below 5%
Microsecond-level grid frequency regulation	Adjustment time of power output	Within 5 milliseconds
Support for coordinated optimization	Reduce the comprehensive cost of distribution network	About 13.2%
Distributed cloud collaborative energy storage system	Reduce peak load of daily operation	11%
Experimental analysis	Increase the rate of absorbing renewable energy sources	14%

In the day-ahead optimization scheduling stage, the experimental data shows that after adopting the AI prediction algorithm, the prediction errors of wind power and photovoltaic power generation have been significantly reduced. This improvement effectively enhances the accuracy of the day-ahead scheduling plan and thereby reduces economic losses caused by prediction deviations. Specifically, as shown in figure 1, the average daily operating cost has decreased by approximately 13.2% compared to the scheduling schemes of case 2, while the active power network loss has also decreased, proving the effectiveness of the multi-objective optimization model in balancing economic benefits and system efficiency.

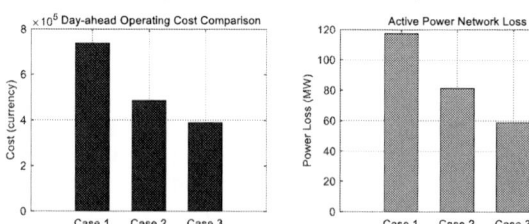

(a) Day-ahead Operating Cost (b) Power Network Loss

Fig. 1 Day-ahead optimization scheduling stage

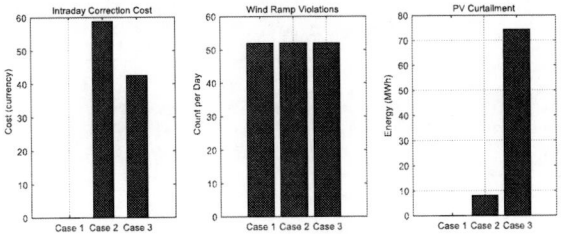

(a) Correction Cost (b) Wind Ramp Violations (c) PV Curtailment

Fig. 2 Intraday optimization scheduling stage

In the intraday optimization scheduling, the AI algorithm demonstrates stronger dynamic adaptability. The rapid response and precise adjustment of real-time data have reduced the correction cost of intraday scheduling by nearly 18.5%, as shown in figure 2. Meanwhile figure 3 gives the energy storage SOC comparison. The algorithm shows good regulation capabilities in handling wind power ramp constraints and photovoltaic curtailment phenomena, significantly alleviating the impact of intermittent renewable energy integration on grid stability. The experiment also reveals that the aggregation management of distributed energy storage further enhances the flexibility of the system, enabling the scheduling strategy to respond more precisely to load fluctuations and changes in renewable energy output.

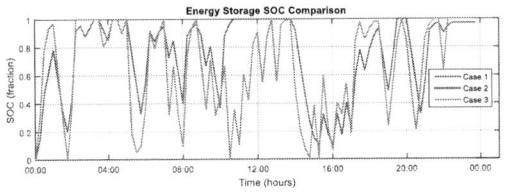

Fig. 3 Energy Storage SOC Comparison

V. SUMMARY

This paper focuses on the research of distributed energy storage and renewable energy coordinated scheduling under multiple time scales, aiming to enhance the economic efficiency and stability of the system through AI optimization algorithms. The study first constructed two-level optimization scheduling models at the day-ahead and intraday levels, respectively addressing the prediction requirements for longer time periods and short-term real-time adjustments. On this basis, an AI-based prediction and scheduling algorithm was designed. The former is used to accurately predict the power generation of renewable energy and load demand, while the latter achieves the optimal allocation of resources through intelligent optimization.

REFERENCES

[1] W. Chen, P. Hu, J. Huang, "Multi-scale climate variations and mechanisms of the onset and withdrawal of the South China Sea summer monsoon." Science China Earth Sciences, vol.65, no.6, pp. 1030–1046, 2022.

[2] Y. Zhao, Z. Wang, D. Luo, C. Chen, B. Ji and G. Li, "Multitimescale Thermal Network Model of Power Devices Based on POD Algorithm." IEEE Transactions on Power Electronics, vol. 39, no. 4, pp. 3906-3924, April 2024.

[3] M. B. Abdelghany, A. Al-Durra, H. H. Zeineldin and F. Gao, "A Coordinated Multitimescale Model Predictive Control for Output Power Smoothing in Hybrid Microgrid Incorporating Hydrogen Energy Storage." IEEE Transactions on Industrial Informatics, vol. 20, no. 9, pp. 10987-11001, Sept. 2024.

[4] C. Wang, S. Ju, J. Sun, et al. "Capsule Neural Network Incorporating Multi-scale Feature Attention and Its Application in Text Classification." Chinese Journal of Information Processing, vol.6, no.1, 65-74, 2022.

[5] Z. Fei, Z. Zhang and K. -L. Tsui, "Deep Learning Powered Online Battery Health Estimation Considering Multitimescale Aging Dynamics and Partial Charging Information." IEEE Transactions on Transportation Electrification, vol. 10, no. 1, pp. 42-54, March 2024.

[6] J. Fan, X. Zhang, Y. Zou and J. He, "Multitimescale Feature Extraction From Multisensor Data Using Deep Neural Network for Battery State-of-Charge and State-of-Health Co-Estimation." IEEE Transactions on Transportation Electrification, vol. 10, no. 3, pp. 5689-5702, Sept. 2024.

[7] F. Liu, D. Yu, W. Su, S. Ma and F. Bu, "Adaptive Multitimescale Joint Estimation Method for SOC and Capacity of Series Battery Pack." IEEE Transactions on Transportation Electrification, vol. 10, no. 2, pp. 4484-4502, June 2024.

[8] M. Wei, J. He, F. Xiong, "Research on Integrated Power Balance Method of Generation, Network, Load and Storage for New Power System." Electrical Engineering Technology, no.18, pp. 34-38, 2022.

[9] B. Liu, J. Zhang, W. Luan, B. Zhao, Z. Liu and Y. Yu, "Adaptive Multitimescale Event Detection in Nonintrusive Load Monitoring Based on Minimum Description Length Principle." IEEE Transactions on Instrumentation and Measurement, vol. 73, pp. 1-14, 2024.

[10] S. Pandey et al., "Multi-Criteria Decision-Making and Robust Optimization Methodology for Generator Sizing of a Microgrid." IEEE Access, vol. 9, pp. 142264-142275, 2021.

2025 International Conference on Advanced Energy Systems and Power Electronics (AESPE 2025)

A Method for Medium Voltage Underground Cables Incipient fault Location Based on the Sheath Grounding Current

Hengzhi Ye
College of Electrical Engineering
Sichuan University
Chengdu China
1294975822@qq.com

Jian Dai
State Grid Wuxi Power Supply Company
State Grid Corporation of China
Wuxi, China

Jun Qin
State Grid Wuxi Power Supply Company
State Grid Corporation of China
Wuxi, China

Wenhai Zhang
College of Electrical Engineering
Sichuan University
Chengdu China

Jinfeng Hu
State Grid Wuxi Power Supply Company
State Grid Corporation of China
Wuxi, China

Lei Su
1 State Grid Wuxi Power Supply Company
State Grid Corporation of China
Wuxi, China

Yuzhe Li
College of Electrical Engineering
Sichuan University
Chengdu China
liyuzhe0124@163.com

Abstract—**With the continuous promotion of urbanization, medium voltage cable is widely used in urban power grid, and the section location of its incipient fault is of great significance to ensure the reliability of power supply. For medium voltage three core cable, a incipient fault section location scheme based on the characteristics of cable sheath grounding current in small current grounding system is proposed in this paper. Firstly, the phase analysis of the the sheath grounding current in the fault and non fault sections under the condition of incipient fault is carried out, and then the section location method based on the phase difference of the first half wave of the the sheath grounding current and the auxiliary location method based on the peak value of the the sheath grounding current are proposed by taking advantage of the characteristics that the phase of the first half wave of the sheath grounding current in the fault section is opposite to that in the non fault section and the peak value is the highest. Finally, simulation software PSCAD/EMTDC is used for simulation verification. Simulation results show that the accuracy of the proposed method is 100% in different fault sections, and it still maintains good accuracy in noisy environment.**

Keywords- Medium voltage underground cable; Incipient fault; Sheath grounding current; Fault section location; Introduction

---

Supported by the Science and Technology Project of State Grid Corporation of China (Project No.: 5400-202318550A-3-2-ZN).

I. INTRODUCTION

With the continuous promotion of urbanization, compared with overhead lines, cables have the advantages of convenient laying, small floor area, high power supply reliability, and small operation and maintenance workload. Therefore, urban core areas, chemical plants, mines and other areas with high requirements for power supply reliability have been widely used. Cables can effectively avoid the potential safety hazards of overhead lines in severe weather such as strong winds and lightning strikes. Therefore, in the construction of power grids in urban areas, medium voltage cables are more and more used. For the purpose of power supply reliability and safety, the section location of incipient fault of medium voltage cable can reduce the manual inspection of fault points, which plays an important role in the actual distribution system.

Most of the cable incipient fault are caused by partial discharge due to partial defects in the cable insulation or cable joints[1-2]. Over time, this kind of fault may occur repeatedly and eventually evolve into a permanent fault[3]. The incipient fault of medium voltage cable usually has the characteristics of short fault time and small current amplitude, so it is difficult to be identified and handled by the line protection system in time.

After finding the incipient fault, it is necessary to locate the incipient fault and identify the cable section with fault through certain methods. At present, the methods can be divided into two categories: active injection method and passive detection method. The passive detection method can be divided into steady-state method and transient method. Active injection method identifies faults through signal excitation. Liu et al. used pulse width modulation rectifier to inject zero sequence current to regulate the neutral point voltage, and locates the fault based on the amplitude difference of feeder zero sequence current[4]. In the research conducted by Ye et al.[5], the low-frequency voltage is applied by flexible switch, and the adaptive criterion is established by using the amplitude ratio of inductance to capacitance current.

In the passive detection method, the steady-state method maps the measurement and control points and lines to the traveling salesman problem. Wei et al. and Zhang et al. used the improved ant colony algorithm[6,7]. Li et al. introduced the pheromone disturbance and global update mechanism[8], and In the research conducted by Liu et al. And Dong et al.[9,10], the fault tolerance through hybrid genetic algorithm and switch function optimization model is improved for FTU alarm distortion. The transient rule focuses on the time-frequency characteristics of faults. Zhang et al. constructed the direction parameter D Based on the reverse polarity of line voltage and zero mode current[11]. Sicong et al. used the similarity difference of zero mode current at both ends of the section to locate[12]. Jiang et al. combined ceemdan decomposition and fuzzy entropy ratio to extract high-frequency features[13]. In the research conducted by Chang et al.[14], the fault section is identified by the positive/negative correlation of the derivative of phase voltage and current abrupt variable. Zhang et al. and Liang et al. analyzed the modulus amplitude and phase characteristics of grounding wire current for cable network[15,16]. The transient method can also be further combined with deep learning and graph structure analysis. Chang et al. proposed the correlation coefficient method based on the phase current fault component, constructed the fault component network through karrenbauer transform, and used the phase current waveform correlation of adjacent intelligent electronic devices (IED) to realize the single-phase fault location of ineffective grounding system, without zero sequence signal and with automatic phase selection[17]. Chen et al. designed drawing attention network (GAT) and consistency risk control (CRC) fusion model, according to the characteristics of distribution network topology dynamic aggregation nodes, combined with CRC quantitative risk prediction, significantly improves the generalization ability and dual fault location accuracy under topology changes. He et al. integrated one-dimensional convolutional neural network (ID-CNN) and spatiotemporal graph convolution network (ASTGCN), extracts the spatiotemporal characteristics of multi-source telemetry data and maps them to the fault branch, and realizes the robust location of high resistance grounding fault through the full connection network, which has strong adaptability to data loss and noise interference[19].

For incipient fault, due to its self recovery characteristics, the active injection method and the method based on the steady-state characteristics of fault current are not suitable for the section location of incipient fault. The use of deep learning method requires a large number of data samples for training, otherwise the problem of insufficient fitting and robustness will occur, and the measured data of incipient fault is small, which may not meet the training needs of sample data.

In view of the above problems, this paper takes the medium voltage three core cable as the research object, and puts forward a new scheme of cable incipient fault section location based on the characteristics of the cable sheath grounding current under small current grounding system. According to the characteristics that the phase of the first half wave of the sheath grounding current in the fault section is opposite to that in the non fault section during the incipient fault, and the current peak is the highest, the section location method based on the phase of the first half wave of the sheath current and the auxiliary location method based on the current peak are proposed. Compared with the traditional method, this method is less affected by other disturbances and noise, simple measurement, small amount of calculation, convenient for real-time section positioning, and the positioning effect is less affected by fault resistance and fault distance.

II. CHARACTERISTIC ANALYSIS OF THE SHEATH GROUNDING CURRENT FOR INCIPIENT FAULT

When the cable is in normal operation, the phase difference of three-phase current in the three-core cable is 120°, and the zero sequence current is zero. Since the sum of three-phase current vectors is almost zero during load balancing, the induced currents cancel each other out, and the grounding current in the shielding layer is zero.

When an incipient fault occurs in a section of cable, the insulation between the sheath and the cable is broken down by an arc, and the cable discharges to the sheath, forming a fault current flowing from the cable to the earth in the sheath.At the same time, due to the electromagnetic coupling between the sheath and the cable core, the zero sequence current generated by the faulty cable forms an induced current in the sheath.The induced current also flows to the earth through the grounding point of the sheath. Therefore, the current flowing to the earth through the sheath can be expressed as the sum of the arc leakage current and the induced current of the sheath. In case of incipient fault, the grounding current of sheath is shown in Figure 1.

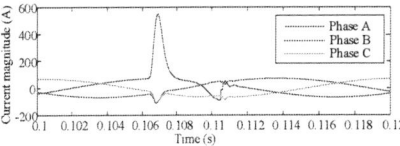

(a) The three-phase current

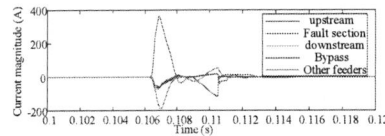

(b) The sheath grounding current

Figure 1. Waveform of three-phase current and the sheath grounding current during incipient fault

979-8-3315-2490-6/25 $31.00 © 2025 IEEE

When incipient fault occurs to the cable, transient arc breakdown occurs between the cable and the sheath, and the fault current flows to the ground in two parts through the grounding points at both ends of the cable sheath. At the same time, due to electromagnetic coupling, the induced current is generated in the sheath and flows to the ground together with the arc current. For the non fault section, although there is no electrical connection between the sheath and the cable core, the induced current will also be generated due to the flow of zero sequence current. In this part, the phase of the sheath grounding current in fault section and non fault section is analyzed.

A. The sheath grounding current in fault section

The total the sheath grounding current is obtained by adding the currents at both ends of the shielding layer. The current flowing from the line to the grounding is defined as the positive direction, and the total the sheath grounding current is:

$$i_g(n) = i_h(n) + i_e(n) \qquad (1)$$

Where, n represents the sampling sequence, $i_g(n)$ represents the sheath grounding current of the cable, $i_h(n)$ represents the sheath head grounding current of the cable, and $i_e(n)$ represents the sheath tail grounding current of the cable.

Because the arc leakage current is much larger than the electromagnetic induction current in the fault section, only the influence of the arc leakage current on the phase of the the sheath grounding current is considered.

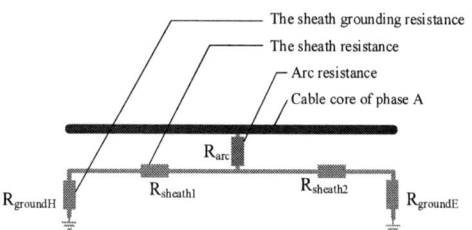

Figure 2. Schematic diagram of fault section in case of incipient fault

As shown in Figure 2, when the incipient fault occurs, the system capacitive current flows to the earth through the arc resistance and the shielding layer grounding resistance. For Ungrounded systems, the voltage at the fault point can be expressed as:

$$\dot{U}_{AD} = \frac{R_{arc} + R_s}{R_{arc} + R_s + Z_{cable}} \dot{E}_A \qquad (2)$$

Where, $\dot{U}_{AD}$ is the residual voltage at the cable fault point, R_{arc} is the fault arc resistance, Z_{cable} is the cable impedance, $\dot{E}_A$ is the no-load phase voltage of phase A, and R_s is the total sheath grounding resistance, and its value is expressed as:

$$R_s = \frac{(R_{sheath1} + R_{groundH})(R_{sheath2} + R_{groundE})}{(R_{sheath1} + R_{groundH}) + (R_{sheath2} + R_{groundE})} \qquad (3)$$

Where, $\dot{E}_B$ and $\dot{E}_C$ are no-load phase voltages of phase B and phase C. Then the zero sequence voltage $\dot{U}_{k0}$ can be obtained as:

$$\dot{U}_{k0} = \frac{1}{3}\left(\dot{U}_{AD} + \dot{U}_{BD} + \dot{U}_{CD}\right)$$

$$= \frac{\left(\dot{U}_{AD} + \dot{E}_B - \left(\dot{E}_A - \dot{U}_{AD}\right) + \dot{E}_C - \left(\dot{E}_A - \dot{U}_{AD}\right)\right)}{3} \qquad (5)$$

$$= -\dot{E}_A + \dot{U}_{AD}$$

Therefore, the capacitive current flowing into the fault point in the non fault phase is:

$$\begin{cases} \dot{I}_B = \dot{U}_{BD} j\omega C_0 \\ \dot{I}_C = \dot{U}_{CD} j\omega C_0 \end{cases} \qquad (6)$$

The current flowing back from the grounding point is:

$$\dot{I}_D = \dot{I}_B + \dot{I}_C \qquad (7)$$

The vector diagram of three-phase voltage and fault current at incipient fault is shown in Figure 3. It is worth noting that due to the arc resistance of incipient fault and the grounding resistance of sheath, there is still residual voltage on phase a of fault phase, and the potential of phase A is not 0 at this time. Therefore, compared with the general short circuit fault, the fault capacitance current in this case is smaller, and the grounding current $\dot{I}_D$ is not completely perpendicular to the zero sequence voltage $\dot{U}_{k0}$, but slightly less than 90° ahead of $\dot{U}_{k0}$. The angle is determined by the arc resistance of incipient fault and the grounding resistance of the sheath.

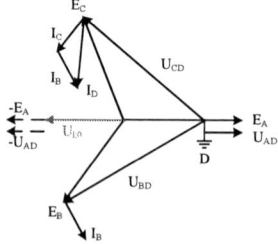

Figure 3. Vector diagram of three-phase voltage and fault current at incipient fault

B. The sheath grounding current in non fault section

When the three-phase currents are symmetrical, the the sheath grounding current is 0. Therefore, for positive sequence or negative sequence current, whether the cable is an independent shielding layer or a common shielding layer, the final the sheath grounding current is offset or directly 0. Therefore, only the influence of zero sequence current on the sheath grounding current is considered next. At this time, the three-phase zero sequence current is:

$$i_0 = I_0 \cos(\omega t + \theta) \qquad (8)$$

Where, I_0 is the amplitude of zero sequence current and θ is the relative phase angle between zero sequence current and positive sequence phase A current. The calculation shows that the induced current of each phase of the three core cable with independent sheath is:

$$i_{gA0} = i_{gB0} = i_{gC0} = \frac{\mu I_0 l}{2\pi R_s} \ln \frac{b}{a} \sin(\omega t + \theta) \qquad (9)$$

C. Section location method based on phase difference of first half wave of the sheath grounding current of cable

In case of incipient fault, zero sequence voltage will be generated at the fault location, and zero sequence current will flow from the fault point to both sides. For the fault section, the direction of zero sequence current flows from the line to the bus or ring main unit; For the non fault section, the direction of zero sequence current flows from the bus to the line, and the phase difference between the two is 180°. It can be seen from the derivation in Chapter II that the phase of the main component of the the sheath grounding current in the fault section is 90° ahead of the zero sequence current , while the main component of the the sheath grounding current in the non fault section is 90° behind the zero sequence current , so the phase difference between the fault section and the the sheath grounding current in the non fault section is 180°. Therefore, it is considered to use the relative phase of the first half wave to locate the fault section.

To calculate the first half wave phase of the sheath grounding current, it is necessary to determine the reference phase. When the incipient fault occurs, the sheath grounding current will be recorded and the subsequent section location will be carried out. If the phase at the beginning of recording is recorded as 0°, the method for calculating the first half wave phase angle of each section of cable is as follows:

The sign function is used to normalize the current amplitude to facilitate the calculation of phase, namely:

$$i_{g,sign}(n) = \begin{cases} 1 & i_g(n) > 0 \\ 0 & i_g(n) = 0 \\ -1 & i_g(n) < 0 \end{cases} \qquad (10)$$

Where, i_{gA0}, i_{gB0} and i_{gC0} are the induced currents caused by zero sequence current in three phases A, B and C respectively, μ Is the magnetic permeability of the cable insulation. Then the total steady-state induced current of the shielding layer in the non fault section can be obtained as:

$$\dot{I}_{ind} = \dot{I}_{gA0} + \dot{I}_{gB0} + \dot{I}_{gC0} = 3\dot{I}_{g0} \qquad (11)$$

Wherein, $\dot{I}_{ind}$ is the steady-state induced current of the sheath and $\dot{I}_{g0}$ is the zero sequence induced current of each phase.

III. LOCATION METHOD OF INCIPIENT FAULT SECTION

Where, n represents the sampling sequence, $i_g(n)$ represents the the sheath grounding current of the cable at the sampling point n, and $i_{g,sign}(n)$ represents the normalized the sheath grounding current.

Then, calculate the phase angle of each section based on the phase at the beginning of recording. Through the first non zero point of the data, the time of the first half wave range phase reference in each section can be obtained. Since the period is 0.02 seconds, the total phase difference of the first half wave at this moment compared with the beginning can be expressed as:

$$\varphi_{all} = \begin{cases} \dfrac{t_{first\_non\_zero}}{0.02} \times 360° & i_{g,sign}\left(t_{first\_non\_zero}\right) = 1 \\ \\ \dfrac{t_{first\_non\_zero}}{0.02} \times 360° + 180° & i_{g,sign}\left(t_{first\_non\_zero}\right) = -1 \end{cases}$$

$$(12)$$

Where, φ_{all} represents the total angle from the sampling start point, $t_{first\_non\_zero}$ represents the time corresponding to the first non zero point of the waveform, and $i_{g,sign}\left(t_{first\_non\_zero}\right)$ represents the amplitude corresponding to the first non zero point of the waveform. When the first non zero amplitude is -1, it is necessary to add 180° to the total angle to indicate that the phase is opposite to that when the amplitude is 1.

Normalize the angle to $[0°, 360°)$:

$$\varphi = \mathrm{mod}(\varphi_{all}, 360) \qquad (13)$$

Where, φ represents the first half wave phase angle normalized to $[0°, 360°)$, and the function $\mathrm{mod}(\varphi_{all}, 360)$ represents the remainder of dividing by 360.

Thus, the phase difference of the sheath grounding current between a certain section and its upper and lower sections can be calculated:

$$\begin{cases} \Delta\varphi_h = |\varphi - \varphi_h| \\ \Delta\varphi_e = |\varphi - \varphi_e| \end{cases} \qquad (14)$$

Where, $\Delta\varphi_h$ represents the phase difference between the section and its upstream section, $\Delta\varphi_e$ represents the phase difference between the section and its downstream section,

φ、φ_h、φ_e represents the first half wave phase angle of the the sheath grounding current of the fault section, upstream section and the downstream section.

In practice and simulation, due to the influence of line reactance, system parameters, fault parameters, etc., the phase of the the sheath grounding current between the fault section and its upstream and downstream sections may not be exactly 180°, so it is necessary to set an interval, and it is considered that the phase difference within the interval meets the condition of opposite phase:

$$\Delta\varphi \in \left[180° - \varphi_{set}, 180° + \varphi_{set}\right] \tag{15}$$

Where, $\Delta\varphi$ is the phase difference with upstream section and the downstream section, φ_{set} is the phase margin, which can be taken as $\varphi_{set} = 15°$ here. If the value of $\Delta\varphi_h$, $\Delta\varphi_e$ both meets this formula, it is considered that incipient fault occurs in this section.

A. Auxiliary section location method based on the peak value of grounding current of cable shielding layer

When the cable section is located at the head or end of the feeder, there is only one adjacent cable section. In case of incipient fault in this section, the phase of the sheath grounding current in two adjacent sections is opposite, and it is impossible to judge which section has fault, so it is necessary to introduce auxiliary positioning method to ensure correct section positioning in case of incipient fault in the first or last section of feeder cable.

After decoupling the transient characteristics of three-phase cables, the γ modulus of current and voltage can be used as the basis for judging whether there is incipient fault, and the γ modulus of fault section is greater than that of non fault section[16]. The current γ modulus is related to the grounding current of the shielding layer, so the peak value of the the sheath grounding current in the fault section is greater than the the sheath grounding current in the non fault section. This can be used as the basis for judging the fault section.

The calculation formula of current peak during incipient fault is as follows:

$$i_{gmax} = \max_{n\in[n_{start}, n_{end}]} \left|i_g(n)\right| \tag{16}$$

Where, i_{gmax} represents the maximum value of all sampling points, n represents the sampling sequence intercepted in the identification process, $i_g(n)$ represents the instantaneous value of the corresponding the sheath grounding current at time n. The start point n_{start} and end point n_{end} of incipient fault represent the start point and end point of the record.

For cables located in the first or last section of the feeder, the incipient fault section identification shall meet the following conditions: 1. the difference between the first half wave phase of the sheath grounding current and adjacent sections during incipient fault is within the threshold; 2. during

incipient fault, the peak value of sheath grounding current is greater than that of adjacent sections. If the above conditions are met, it is considered that incipient fault has occurred in this section.

B. Location process of cable fault section in case of incipient fault

Based on the above positioning method of cable fault section based on the peak value and effective value of shielding layer grounding current, the specific process of monitoring and positioning of cable incipient fault is proposed, as shown in Figure 4.

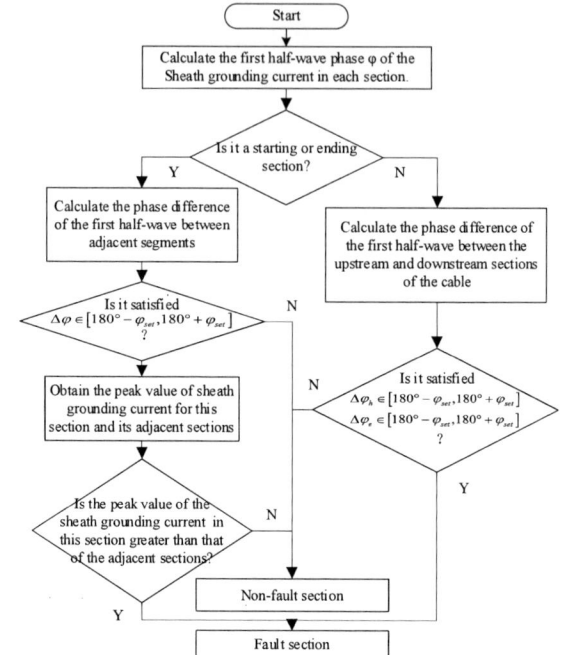

Figure 4. Flow chart of cable early grounding fault section location

IV. SIMULATION AND VERIFICATION OF CABLE INCIPIENT FAULT LOCATION METHOD BASED ON PSCAD/EMTDC

The simulation analysis is carried out in the simulation software PSCAD v4.6. In terms of hardware, the computer used in the simulation uses a CPU of AMD Ryzen 7 6800H and a graphics card of NVIDIA RTX 3050 Ti. Kilicay model is adopted for the arc resistance model of incipient fault, and the arc resistance is set according to this model]. The three core cable adopts an independent metal shielding layer model, with small resistance grounding at both ends. The simulation time is 0.3s, the solution time step is 10us, and the sampling rate is 10kHz.

After the simulation, Matlab is used to process the waveform and observe and compare it. The schematic diagram of the simulation model is shown in Figure 5. The other overhead lines are equivalent simulated by lumped parameters. The length parameter settings of each section of the cable are shown in Table 1.

TABLE I.	THE LENGTH OF EACH CABLE SECTION
Segment Serial Number	Segment Length/km
C1	6
C2	6
C3	6
C4	6
C5	20

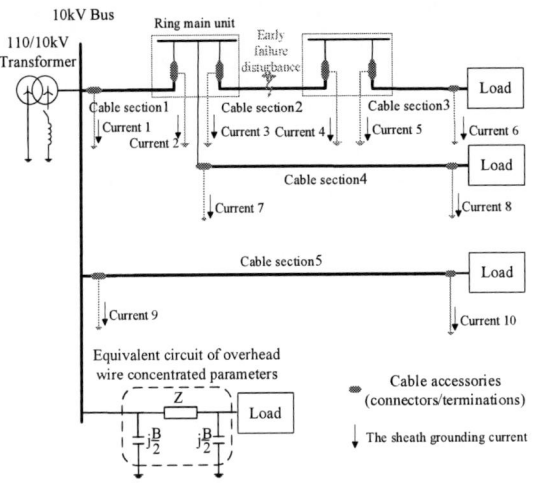

Figure 5. Schematic diagram of PSCAD simulation model

Incipient faults are set in different cable sections. In the incipient fault of each cable section, the fault durations from 0.005 seconds to 0.08 seconds and different closing anglesset are set to obtain 110 groups of data.A total of 550 groups of data are simulated and the incipient fault section location method is realized by MATLAB software. The results of section location method are shown in Table 2.

TABLE II.	THE RESULT OF THE SECTION LOCATION METHOD		
Fault section	Number of Simulations	Successful positioning count	Accuracy Rate(%)
C1	110	110	100
C2	110	110	100
C3	110	110	100
C4	110	110	100
C5	110	110	100

Considering the impact of noise on section positioning, it is necessary to ensure that the noise will not interfere with the determination of the first half wave phase. To avoid interference in phase calculations, samples below the threshold are set to zero. This can suppresses noise from normal operations or disturbances:

$$i_g(n) = \begin{cases} i_g(n) & i_g(n) \ge i_{\varphi set} \\ 0 & i_g(n) < i_{\varphi set} \end{cases} \quad (17)$$

Where, n represents the sampling sequence, $i_g(n)$ represents the the sheath grounding current of the cable at the sampling point n, $i_{\varphi set}$ represents the set zero threshold, and the threshold setting should at least avoid the maximum the sheath grounding current and noise that may occur during normal operation. Then the results of section positioning method under noise environment are shown in Table 3. It can be seen that when the signal-to-noise ratio is large, the section positioning results are basically not affected, and when the signal-to-noise ratio continues to increase to a certain level, the section positioning accuracy will be greatly reduced. Therefore, this method is suitable for the environment with large signal-to-noise ratio, or for the section location after noise reduction when the signal-to-noise ratio is small.

TABLE III.	RESULTS OF SECTION LOCATION METHOD UNDER THE INFLUENCE OF NOISE		
Signal-to-noise ratio（dB）	Number of Simulations	Successful positioning count	Accuracy Rate（%）
50	550	550	100
40	550	550	100
30	550	550	100
20	550	548	99.64
10	550	162	29.45

V. CONCLUSION

This paper proposes a method of incipient fault section location based on the sheath grounding current. Firstly, the the sheath grounding current of fault section and non fault section in incipient fault is analyzed and the conclusion that their phases are opposite is drawn. Then, based on the characteristics of the sheath grounding current during incipient fault, the method uses the first half wave phase difference to locate the fault section, and uses the peak value of the sheath grounding current for auxiliary location. Finally, the simulation software PSCAD/EMTDC is used to build the equivalent model of cable incipient fault, and the proposed method is verified, which proves that the proposed method can effectively realize the section location of incipient fault, and has good anti-interference ability to noise. Compared with other methods, this method is less affected by other interference and noise. This method is also simple to measure and does not need to calculate and set the threshold.

However, this method still has limitations and needs to be improved. In this method, distributed generation and ring network are not considered, which may make this method not applicable to these situations. The simulation in this paper is only carried out in a relatively simple distribution network. When facing the distribution network with more complex topology, its feasibility still needs to be further verified.

REFERENCES

[1] He, J., He, K., & Dong, B., "Electric field characteristics investigation and electrical tree propagation of distributed network cable joint with defects," Electric Power Engineering Technology, 2023, 42 (01), 154-161.

[2] Yu, Z., Xu, Y., Liu, Y., Yan, Y., Jiang, X., & Wang, K., "Analysis of Characteristic Quantity of Incipient Fault of Medium Voltage Cable Joints in Water-Logged Environment Based on Fluid

979-8-3315-2490-6/25 $31.00 © 2025 IEEE

Characteristics," IEEE Transactions on Instrumentation and Measurement, 2024, 73.

[3] Yu, Z., Xu, Y., Qian, Q., Duan, C., & Jiang, X., "Evolution Process and Characteristics of Medium Voltage Cable Joints' Incipient Faults in Humid Environment," Gaodianya Jishu/High Voltage Engineering, 50(3), 2024, 1311–1321.

[4] Liu, S., Yu, K., Zeng, X., Liu, Z., & Chen, X., "Fault location method of a non-effective earthed system based on zero sequence current amplitude continuous regulation," Dianli Xitong Baohu Yu Kongzhi/Power System Protection and Control, 2021, 49(9), 48‑56.

[5] Ye, Y., Ma, X., Lin, X., Li, Z., Xu, F., Wang, C., Ni, X., & Ding, C., "Active Fault Locating Method Based on SOP for Single Phase Grounding Faults in the Resonant Grounding Distribution Network," Zhongguo Dianji Gongcheng Xuebao/Proceedings of the Chinese Society of Electrical Engineering, 2020, 40(5), 1453‑1464.

[6] Wei Q., Mingna J., Wei W., "Application of the Improved Ant Colony Algorithm to Distribution Network Fault Location," Shandong Electric Power, 2017.

[7] Zhang, Y., Zhou, R., & Zhong, K., "Application of improved ant colony algorithm in fault-section location of complex distribution network," Dianwang Jishu/Power System Technology, 35(1), 2011, 224‑228.

[8] Li, H, Zhu,Y, & Ma,H., "Least Squares Support Vector Machine Based Incipient Fault Identification in Non-solidly Grounding System," Journal of Electrical Engineering, 2023, 18(03), 297-306.

[9] Liu, L., "Automatic Localization of Fault Section in Active Distribution Networks Based on Hybrid Genetic Algorithm," Electric Switchgear, 2024, 62 (03), 78-81.

[10] Dong, X., Zhu, Y., & Wang, Q., "Regional fault location and fault tolerance analysis of distribution network based on genetic algorithm," Chinese Journal of Construction Machinery, 2023, 21 (06), 607-612.

[11] Zhang, L., Xu, B., Xue, Y., & Gao, H., "Transient fault locating method based on line voltage and zero-mode current in non-solidly earthed network," Zhongguo Dianji Gongcheng Xuebao/Proceedings of the Chinese Society of Electrical Engineering, 2012,m32(13), 110‑115.

[12] Shicong M A , Bingyin X U , Houlei G ,et al, "An Earth Fault Locating Method in Feeder Automation System by Examining Correlation of Transient Zero Mode Currents," Automation of Electric Power Systems, 2008.

[13] Jiang, H., Li, X., Liu, X., You, J., Wang, X., Fang, Y., & Sheng, X., "Fault Section Location Method for Multi-branch Distribution Lines Based on Dual Fuzzy Entropy Criterion," Smart Power, 2025, 53(1), 90-97.

[14] Chang, Z., Song, G., Huang, W., Guo, S., & Zhang, W., "Phase Voltage and Current Fault Components Based Fault Segment Location Method Under Single-Phase Earth Fault in Distribution Network," Dianwang Jishu/Power System Technology, 2017, 41(7), 2363-2369.

[15] Zhang, P., Kong, L., Liang, R., Xu, B., & Peng, N., "Fault Location Method for Three-Core Cable Using Amplitude Ratio of Shield-Grounding Wire Currents," IEEE Transactions on Industrial Informatics, 2023, 19(6), 7456-7467.

[16] Liang, R., Peng, N., Zhang, Z., Li, J., Kong, L., & Wang, Q., "Single-phase Grounding Fault Section Location of the Distribution Cable Networks Based on the Time-frequency Analysis of the Transient Featured Moduli," Zhongguo Dianji Gongcheng Xuebao/Proceedings of the Chinese Society of Electrical Engineering, 2023, 43(23), 9098-9113.

[17] Z. Chang et al, "Phase Current Based Fault Section Location for Single-Phase Grounding Fault in Non-Effectively Grounded Distribution Network," IEEE Transactions on Industry Applications, 2025, vol. 61, no. 3, pp. 5242-5250.

[18] Chen, X., Sun, L., Li, Y., Li, B., Wang, L., & Cai, Y., "A Fault Section Location Method Based on Graph Attention Network and Conformal Risk Control in Distribution Network," Dianwang Jishu/Power System Technology, 2023, 47(12), 4866-4876.

[19] He, X., Gao, H., Huang, Y., Gao, Y., Wang, R., & Liu, J., "Fault section location for a distribution network based on one-dimensional convolution and graph neural networks," Dianli Xitong Baohu Yu Kongzhi/Power System Protection and Control, 2024, 52(17), 27‑39.

2025 International Conference on Advanced Energy Systems and Power Electronics (AESPE 2025)

Research on Low-Inductance Packaging Design and Switching Characteristics of GaN HEMT Half-Bridge Modules for High-Frequency Converters

Xiangqi Qiu
School of Mechanical and Electrical Engineering
Guilin University of Electronic Technology
Guangxi, China

Xiao Zhang
School of Mechanical and Electrical Engineering
Guilin University of Electronic Technology
Guangxi, China

Pengfei Lu
School of Mechanical and Electrical Engineering
Guilin University of Electronic Technology
Guangxi, China

Song Wei *
School of Mechanical and Electrical Engineering
Guilin University of Electronic Technology
Guangxi, China
swei2020@126.com

Abstract—A low-parasitic-inductance gallium nitride (GaN) high-electron-mobility transistor (HEMT) half-bridge integrated power module is proposed in this paper. Organic substrate packaging technology is adopted, and the power loop inductance is reduced to 457.07 pH by optimizing the layout and packaging structure to meet the demand for low parasitic inductance in high-frequency power conversion systems. Combined with Ansys Q3D electromagnetic simulation and double-pulse testing, the turn-on and turn-off inductances of the driving loop are 1.56 nH and 1.33 nH. Under the 80V/16A operating condition, the voltage overshoot of the module is only 3.39V. Comparative experiments have shown that compared with commercial modules, the designed low-inductance module exhibits significant advantages in efficiency, stability, and dynamic response. Through multi-physics field collaborative optimization, this study solves the problems of voltage oscillation and loss caused by parasitic inductance in high-frequency applications of the module. It provides important theoretical guidance and practical references for low-inductance packaging design, board-level structure optimization, and high-frequency module integrated packaging of GaN HEMT half-bridge modules.

Keywords-GaN HEMT; integrated power module; parasitic inductance; switching characteristics; Double-Pulse Test

I. INTRODUCTION

As a third-generation wide-bandgap semiconductor material, gallium nitride (GaN) demonstrates distinct advantages over conventional silicon and silicon carbide (SiC) materials with respect to bandgap width, breakdown field strength, electron mobility, and saturation drift velocity. Such attributes establish GaN's technological preeminence in the domain of high-frequency power conversion, providing a robust material foundation for the realization of high-power-density power conversion systems.

Packaging technology significantly impacts the high-frequency performance of devices. Conventional TO-type packages, which utilize wire bonding for chip-to-external-circuit connections, exhibit source inductance of approximately 5-15 nH and total gate-loop inductance reaching

10-20 nH. Such high parasitic inductance limits switching speed [1]. Infineon's CoolGaN™ G3 series adopts RQFN packaging, achieving a 30% reduction in parasitic inductance [2]. ROHM's specially designed QFN package for multi-chip modules reduces parasitic inductance to 0.5 nH through three dimension (3D) stacked configuration [3]. The research team led by Prof. Chen Jing at Hong Kong University of Science and Technology proposed a 3D stacked packaging solution for GaN/SiC cascode devices, effectively minimizing parasitic inductance by shortening current pathways [4].

Despite the high-frequency characteristics of GaN HEMTs facilitating significant reductions in the volume and weight of power converters while enhancing the overall performance of the system, critical challenges are encountered in practical implementation. Voltage overshoot, ringing phenomena, and electromagnetic interference induced by packaging parasitic inductance under high-frequency operation may lead to a reduction in efficiency of up to 15% and pose risks of device failure [5]. Therefore, this study aims to minimize parasitic inductance by optimizing the packaging structure and organic substrate layout. It delves into the impact of packaging structures on device switching performance. The research encompasses innovative packaging design, precise parasitic inductance modeling and simulation, and actual device fabrication. It strives to offer a high - performance, high - reliability GaN HEMT half - bridge integrated module solution for high - frequency converter applications. This will foster the broad application of wide - bandgap semiconductor technology in power electronics.

II. LOW INDUCTANCE PACKAGE DESIGN

A. Structure of the Low-Inductance Packaging Design

The low-inductance GaN half-bridge integrated power module is designed to achieve high-efficiency, high-frequency, and high-power-density power conversion. Its core lies in packaging two GaN chips, one driving chip, and a bus capacitor within a single module, thereby simplifying circuit design and reducing parasitic inductance. This design

979-8-3315-2490-6/25 $31.00 © 2025 IEEE

minimizes the substrate layout area [6]. A schematic diagram of the packaging structure is shown in Fig. 1.

Fig. 2 depicts the package circuit. At its core are two 100 V/60 A enhanced-mode GaN HEMTs and a GaN half-bridge driver chip. The module incorporates bus capacitance, turn-on and turn-off resistors for the upper and lower arms, bootstrap capacitance, and decoupling capacitance for the driver chip's power supply. With a turn-on resistance of 10Ω and a turn-off resistance of 2Ω, the turn-on resistance is designed for flexibility. It can be adjusted by connecting external resistors in parallel via external pins, allowing for precise control of the turn-on speed. The integrated driver design eliminates the need for external driver circuits, reduces parasitic parameters, minimizes voltage and current spikes [7].

Low-inductance GaN half-bridge integrated rate module selected 0.6 mm thick four-layer Flame-Retardant 4 organic substrate, the inner and outer layers of the 1 ounce thickness of copper foil wiring, the GaN chip, driver chip and related components integrated on the substrate. The substrate layout design of the integrated power module is shown in Fig. 3.

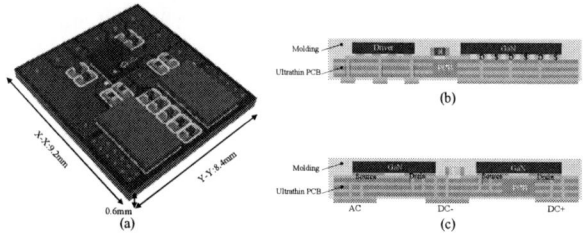

Figure 1.   Schematic of the package structure. (a) 3D model. (b) X-X cross-section. (c) Y-Y cross-section

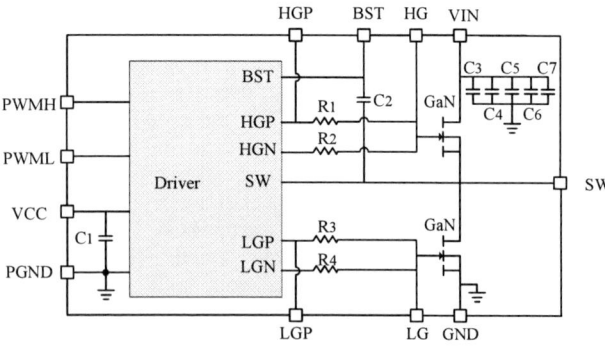

Figure 2.   Power module package circuit schematic.

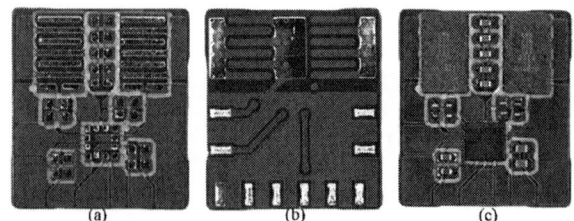

Figure 3.   Substrate Layout Design. (a) Top view. (b) Bottom view. (c) physical modulet.Parasitic Inductance Extraction

B.   Parasitic Inductance of the Low-Inductance Packaging

Parasitic inductance in low-inductance GaN half-bridge integrated power modules primarily stems from internal wiring, connections, and packaging structures. Fig. 4 depicts the simplified parasitic inductance distribution within such modules.

Using Ansys Q3D Extractor to quantify gate and power loop parasitic inductance parameters [8]. Fig. 5 demonstrates that gate loop parasitic inductance exhibits a decreasing trend with increasing frequency, reaching stabilization at 1 MHz.

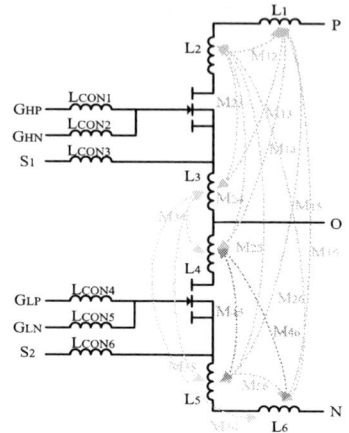

Figure 4.   Parasitic inductance equivalent circuit diagram.

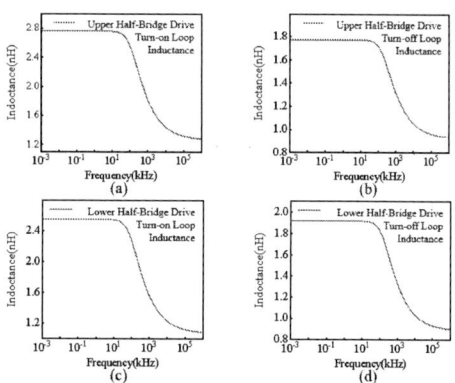

Figure 5.   Inductance curve of driving circuit. (a) Upper half-bridge turn-on circuit. (b) Upper half-bridge turn-off circuit.(c) Lower half-bridge turn-on circuit. (d) Lower half-bridge turn-off circuit.

The module achieves significant total inductance reduction through interleaved internal trace and via routing, leveraging mutual inductance effects [9]. Accounting for mutual inductance, parasitic inductance values at 1 MHz were designated as the observation reference. Table 1 presents the resulting gate loop parasitic inductance distribution.

In (1) parasitic inductance matrix, the off-diagonal elements indicate the mutual inductance, where negative values indicate the mutual inductance cancellation phenomenon between two inductors, while positive values indicate the mutual inductance enhancement. Combined with the detailed analysis and accurate calculation of the simulation data of the parasitic inductance equivalent circuit of the low-inductance GaN half-bridge integrated power module in Fig. 4, the parasitic inductance of the overall power converter circuit is obtained as 457.07 pH.

TABLE I. GaN MODULE DRIVE LOOP PARASITIC INDUCTANCE VALUES

Parasitic Inductance Definition	Define	Value (nH)
$L_{CON1}+L_{CON3}$	Upper bridge opening circuit	1.816
$L_{CON2}+L_{CON3}$	Upper bridge shutdown circuit	1.317
$L_{CON4}+L_{CON6}$	Lower bridge opening circuit	1.562
$L_{CON5}+L_{CON6}$	Lower bridge shutdown circuit	1.334

$$L_S = \begin{bmatrix} L_1 & M_{12} & M_{13} & M_{14} & M_{15} & M_{16} \\ M_{12} & L_2 & M_{23} & M_{24} & M_{25} & M_{26} \\ M_{13} & M_{23} & L_3 & M_{34} & M_{35} & M_{36} \\ M_{14} & M_{24} & M_{34} & L_4 & M_{45} & M_{46} \\ M_{15} & M_{25} & M_{35} & M_{45} & L_5 & M_{56} \\ M_{16} & M_{26} & M_{36} & M_{46} & M_{56} & L_6 \end{bmatrix} \quad (1)$$

III. SWITCHING CHARACTERISTICS ANALYSIS AND TESTING

A. Transient Switching Characteristics Analysis

During the switching transient of the module, voltage overshoot is induced by parasitic inductance within the commutation loop. This phenomenon causes the gate-source voltage of the device to exceed its rated withstand voltage, consequently precipitating device failure. To comprehensively evaluate the switching characteristics of the low-inductance GaN half-bridge integrated power module, a double-pulse test simulation methodology was implemented utilizing LTspice. The resultant double-pulse simulation waveform is illustrated in Fig. 6, with corresponding device selection criteria and parametric configuration detailed in Table 2 [10].

Systematic evaluation of module switching characteristics was performed across a bus voltage range of 60-80 V, with particular emphasis on analyzing voltage/current oscillation dynamics and voltage overshoot during turn-off transients. The experimental results, as illustrated in Fig. 7, present the turn-off voltage and current characteristics of the low-inductance GaN half-bridge integrated power module.

The turn-off loss of the power module can be calculated using (2). Table 3 provides a detailed summary of the results for the maximum voltage and turn-off energy E_{off} under different test conditions [11]. The simulation results demonstrate that under the 80 V/16 A condition, the low-inductance GaN half-bridge integrated power module exhibits a voltage overshoot of 6.72 V and a turn-off loss of approximately 2.08 μJ. Based on this analysis, it can be concluded that the designed low-inductance GaN half-bridge integrated power module exhibits excellent switching characteristics, meeting the stringent requirements of high-frequency power conversion applications.

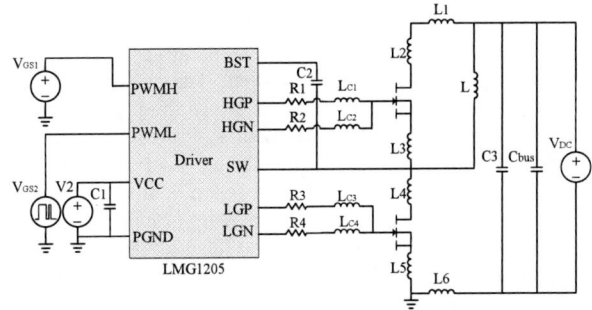

Figure 6. LTspice Double Pulse Simulation Test Diagram.

TABLE II. LTSPICE DOUBLE PULSE SIMULATION TEST DEFINITIONS.

Define	define	value	define	value
Power Loop	V_{DC}	60-80V	L	12uH
	C_{BUS}	60uF	C_3	600pF
	L_1	59.95pH	L_2	25.42pH
	L_3	188.21pH	L_4	98.75pH
	L_5	36.04pH	L_6	49.1pH
Drive Loop（LMG1205）	V_2	5V	V_3	0V
	C_1	1uF	C_2	0.1uF
	R_1、R_3	10Ω	R_2、R_4	2Ω
	L_{C1}	1.816nH	L_{C2}	1.317nH
	L_{C3}	1.562nH	L_{C4}	1.334nH
GaN（INN100W032A）	V_{DS}	60-80V	I_D	10-30A
	V_{GS1}	0V	V_{GS2}	PWM

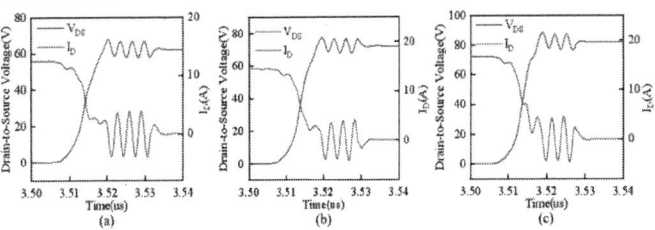

Figure 7. LTspice-simulated turn-off waveforms under Different Bus Voltages. (a) 60 V. (b)70 V. (c) 80 V.

Figure 8. SIMULATION RESULTS OF DOUBLE-PULSE TEST IN LTSPICE.

Bus Voltage (V)	Current (A)	Voltage Overshoot (V)	E_{off} (uJ)
60	12.42	5.56	1.35
70	14.53	5.64	1.58
80	16.67	6.72	2.08

$$E_{off} = \int_{t1}^{t2} V_{ds}(t) \times I_d(t) d_t \quad (2)$$

979-8-3315-2490-6/25 $31.00 © 2025 IEEE

B. Double-Pulse Test Comparison and Validation

This chapter presents a comparative analysis of the switching characteristics between the self-designed low-inductance GaN half-bridge module and a commercially available module with equivalent specifications, based on a standardized testing platform. To eliminate the influence of parasitic parameters from the test board, a unified double-pulse test setup and identical testing conditions were employed. Under a direct current (DC) bus voltage of 80 V, the switching waveforms of both the low-inductance GaN half-bridge integrated power module and the commercial integrated module were measured, with the results illustrated in Fig. 8.

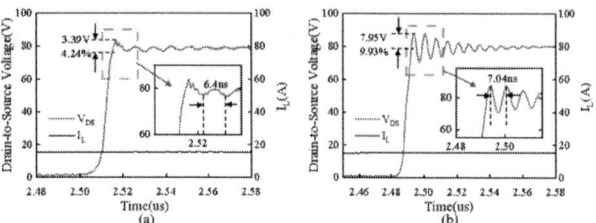

Figure 9. turn-off waveform with bus voltage of 80 V.(a) low-inductance GaN power module. (b) commercial GaN power module

When the bus voltage is 80V, the commercial integrated GaN module exhibits a voltage overshoot of 7.95V, accounting for 9.93% of the bus voltage. In comparison, the low-inductance GaN half-bridge integrated power module demonstrates a voltage overshoot of only 3.39V, representing 4.24% of the bus voltage 5.58% lower than its commercial counterpart. Furthermore, the designed module shows an improved turn-off waveform oscillation period of 6.4ns, compared to 7.04ns for the commercial module. Based on the relationship between voltage oscillation period during turn-off and parasitic inductance, the parasitic inductance of the power module's loop can be calculated using (3) [12].

$$T_{\text{ring}} = 2\pi \cdot \sqrt{L_{\text{Loop}} \cdot (C_{\text{OSS}} + C_{\text{Para}} + C_{\text{Probe}})} \qquad (3)$$

Where T_{ring} represents the oscillation period during turn-off, L_{Loop} denotes the parasitic inductance of the power loop, C_{oss} is the output capacitance of the GaN chip, C_{para} indicates the combined parasitic capacitance of the power module and test board, C_{probe} and signifies the parasitic capacitance of the passive probe.

According to the GaN chip datasheet, the output capacitance measures 450 pF at a drain-source voltage of 80 V. ANSYS Q3D simulations reveal parasitic capacitances of 11.35 pF for the module and 1.19 pF for the test board, respectively. The passive probe contributes an additional 3.9 pF parasitic capacitance. Based on these parameters, the calculated parasitic inductance values are 2.22 nH for the low-inductance GaN half-bridge integrated power module and 2.69 nH for the commercial GaN integrated module.

IV. Conclusions

This paper develops a low-parasitic-inductance GaN HEMT half-bridge integrated module for high-frequency power conversion requirements.The research mainly includes module design, simulation verification, and switching characteristic testing. Using Q3D electromagnetic simulation to optimize the power module structure, experimental data confirms the module's significant advantages in voltage stress suppression, switching loss optimization, and system stability. Efficiency comparison analysis reveals that under high output power conditions, the designed low-inductance power module exhibits notably lower energy loss and higher efficiency, demonstrating clear performance superiority. These research outcomes provide important technical references for packaging design and system integration of high-frequency, high-power GaN devices.

Although experimental results indicate that this design performs excellently in reducing parasitic inductance and increasing power density, there remains room for optimization in thermal management, packaging structure optimization, and fabrication processes. Since gallium nitride generates substantial heat during high-frequency operation, future research could consider optimizing heat dissipation structures and adopting new thermal interface materials like liquid metals to enhance thermal performance. The through-hole substrate selected for the module limits overall structural layout, while blind/buried via designs could provide greater structural optimization potential. Furthermore, the current resin-filled plating for organic substrate vias could be upgraded to copper paste filling or copper plating to minimize parasitic parameters and improve thermal performance to the greatest extent.

Acknowledgment

The information, work, presented herein was supported by Key Technology Research Project of Shaoxing City (Grant No. 2024B11013), Guangxi Key Laboratory of Manufacturing System & Advanced Manufacturing Technology (Grant No. 22-35-4-S018), National Natural Science Foundation of China (Grant No. 52105327).

References

[1] Huang X, Liu T, Li B, et al. Evaluation and applications of 600V/650V enhancement-mode GaN devices[C]//2015 IEEE 3rd Workshop on Wide Bandgap Power Devices and Applications (WiPDA). 2015: 113-118.

[2] "GaN Transistors Fit Standard Si Packages." (2025). [Online]. Available: https://www.edn.com/gan-transistors-fit-standard-si-packages/

[3] Infineon Technologies AG, "Infineon introduces CoolGaN™ G3 Transistor in RQFN," 2025. [Online]. Available: https://www.infineon.com/cms/en/about-infineon/press/market-news

[4] Shu, Ji, et al. "Stacked Strongly Coupled GaN/SiC Cascode Device With Fast Switching and Reclaimed Strong dv/dt Control." IEEE Transactions on Electron Devices (2025).

[5] Feng JJ, Fan YM, Fang D, et al. Research Progress on Packaging Technologies for Gallium Nitride Power Electronic Devices[J]. Journal of Synthetic Crystals, 2022, 51(4): 730-749.

[6] Ke H, Mehrotra U, Hopkins D C. 3-D Prismatic Packaging Methodologies for Wide Band Gap Power Electronics Modules[J]. IEEE Transactions on Power Electronics, 2021, 36(11): 13057-13066.

[7] Wang Q, Yuan JS, Zhao QM. Inductance Calculation of Submarine DC Transmission Lines Based on Finite Element Analysis[J]. Chinese Journal of Ship Research, 2018, 13(1): 114-119.

979-8-3315-2490-6/25 $31.00 © 2025 IEEE

[8] Chen H. Research on Packaging Technology and Switching Characteristics of Low - Inductance Laminated Circuit Power Modules[D]. Guilin University of Electronic Technology, 2024.

[9] Zhang X, Gan Y, Zhang T, et al. Dynamic Current Balancing Optimization of Cu Clip-Bonded SiC power module Based on Layout-Dominated Parasitic Inductance[C]//PCIM Asia 2024; International Exhibition and Conference for Power Electronics, Intelligent Motion, Renewable Energy and Energy Management. 2024: 106-110.

[10] Li H, Gao Z, Chen R, et al. Improved Double Pulse Test for Accurate Dynamic Characterization of Medium Voltage SiC Devices[J]. IEEE Transactions on Power Electronics, 2023, 38(2): 1779-1790.

[11] Wei, W. W. Analysis and Application Research on Switching Characteristics of GaN Devices. [D] Anhui University of Technology, 2019.

[12] Qi Z, Pei Y, Wang L, et al. A Highly Integrated PCB Embedded GaN Full-Bridge Module With Ultralow Parasitic Inductance[J]. IEEE Transactions on Power Electronics, 2022, 37(4): 4161-4173.

Particle Swarm Optimization-Based Microgrid Dispatch with Renewable Energy Integration

Qianlong Li
Tongji University, Shanghai, China
e-mail: keats409@sina.com

Abstract: **The inherent output variability of renewable energy sources poses challenges for integrating wind and photovoltaic generation into power grids. Modern electrical infrastructure increasingly incorporates substantial wind and solar photovoltaic resources. Energy storage systems, with their bidirectional charging capabilities, offer effective peak regulation for microgrids. However, coordinating multiple power sources remains operationally challenging. This study presents an optimized scheduling framework for microgrids integrating wind, solar photovoltaic, micro gas turbines, and energy storage units. The model minimizes operational costs while ensuring efficient system performance. We employ swarm intelligence optimization for model resolution. Case study results validate the model's effectiveness in leveraging diverse resource regulation capabilities. The proposed approach reduces renewable energy curtailment, enhances microgrid economic performance, and significantly improves peak load regulation. These findings demonstrate the methodology's efficacy and robustness.**

Keywords: microgrid; economic dispatch; particle swarm optimization

I. INTRODUCTION

Extensive reliance on hydrocarbon-based energy has resulted in severe ecological degradation and atmospheric pollution. China's pursuit of carbon neutrality and peak carbon emissions has accelerated the transition toward diversified and cleaner energy sources. Microgrids, as key infrastructure for integrating renewable resources such as wind and solar power, are becoming essential components of China's renewable-energy-based power system. While increased wind and photovoltaic integration in microgrids has mitigated environmental pollution, the inherent output variability of these sources presents significant challenges for optimal dispatch. Consequently, researchers are increasingly focusing on microgrid optimization to enhance economic efficiency through strategic scheduling of distributed energy resources.

For the optimal operation of microgrids, numerous scholars have designed various optimization strategies. Wen et al. proposed a multi-microgrid shared energy storage operation optimization strategy based on mixed game theory. The approach establishes a Stackelberg game model with shared energy storage operators as leaders and multi-microgrids as followers, combined with a generalized Nash bargaining model considering compensation mechanisms, to coordinate stakeholder decisions for improving renewable energy utilization and reducing system operating costs[1].Peter et al. proposed a hybrid energy management framework integrating the One-to-One-Based Optimizer (OOBO), K-means clustering algorithm, and Artificial Neural Networks for optimizing energy scheduling and load forecasting in grid-connected microgrids, achieving significant reductions in operational costs and carbon emissions[2].Chen et al. proposed a quantized distributed economic dispatch algorithm for microgrids. The algorithm integrates a dynamic quantization scheme with the Paillier encryption-decryption framework. It achieves exact convergence to the optimal solution while reducing communication overhead and ensuring secure transmission of power-sensitive information[3].Wang et al. proposed a distributed multi-agent reinforcement learning framework for dynamic energy dispatch in microgrids. The approach formulates the problem as a partially observable Markov decision process and employs LSTM networks for state prediction. Each generator operates as an individual agent using multipliers splitting methodology to minimize total generation costs while maintaining power balance and ensuring data privacy[4].Li et al. proposed a distributionally robust chance-constrained negative-emission optimal energy scheduling scheme for off-grid integrated electricity-heat microgrids. The approach achieves negative emission targets through post-combustion carbon capture systems and direct air capture technologies, while employing distributionally robust chance constraints to handle uncertainties in wind power and carbon emissions[5].Zou et al. proposed a risk-averse transactive energy management method for multi-energy microgrids. The method formulates a bi-level optimization model based on Stackelberg game theory. This paper uses an adaptive stochastic optimization approach that coordinates day-ahead and intra-day decisions. A conditional value-at-risk (CVaR) measure captures the operator's risk aversion[6].

This research addresses the dual objectives of minimizing economic costs and environmental impacts while ensuring reliable microgrid operation. We develop a multi-objective optimization framework for scheduling microgrids comprising wind turbines (WT), photovoltaic panels (PV), micro-turbines (MT), and battery storage (BT). An enhanced particle swarm optimization algorithm is employed to solve this complex problem.Results demonstrate that our framework effectively exploits the flexibility of diverse energy resources. The approach substantially reduces renewable energy curtailment, improves microgrid economic performance, and enhances peak load management capability. These findings validate the methodology's robustness and effectiveness.

II. MICROGRID OPTIMAL DISPATCH MODEL

This research presents a cost-optimal operational strategy for microgrids integrating wind turbines, photovoltaic arrays, micro gas turbines, and battery storage, as illustrated in Figure 1. The approach maximizes renewable utilization through economic dispatch and time-of-use (TOU) pricing. An optimization algorithm determines optimal power allocation among generation units, minimizing operational costs while satisfying power balance and system constraints.

979-8-3315-2490-6/25 $31.00 © 2025 IEEE

The proposed strategy employs four key principles to enhance microgrid economic performance:(1) Renewable priority: Wind and photovoltaic generation satisfy load demands preferentially, minimizing reliance on conventional sources.(2) Economic dispatch: When renewable generation is insufficient, the algorithm compares costs among micro gas turbines, grid purchases, and battery discharge to select the most economical option.(3) Surplus management: Excess renewable energy is either sold to the grid or stored in batteries based on economic merit.(4) TOU optimization: Battery charging occurs during low-price periods, while discharging and grid sales are scheduled during peak pricing to maximize revenue.

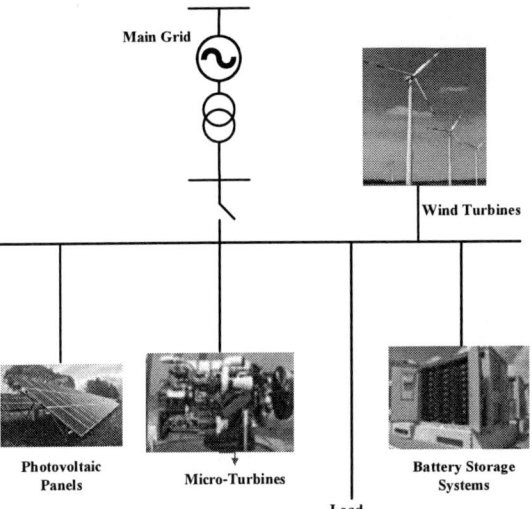

Figure 1. Microgrid System Architecture

A. Micro Gas Turbine Model

Micro gas turbines are low-capacity generation units whose efficiency varies with power output according to：

$$\eta_{\text{MT}} = 0.1068 + 0.4174 \cdot \frac{P_{\text{MT}}}{65} - 0.3095 \cdot \left(\frac{P_{\text{MT}}}{65}\right)^2 + 0.0753 \cdot \left(\frac{P_{\text{MT}}}{65}\right)^3 \quad (1)$$

Where η_{MT} is the turbine efficiency and P_{MT} is the generated power (kW) at time interval t.

The fuel consumption cost is：

$$C_{MT} = C_{ng} \cdot \frac{P_{MT} \cdot \Delta t}{\eta_{MT} \cdot Q_{ng}} \quad (2)$$

The parameter C_{MT} indicates the fuel expenditure associated with operating the micro gas turbine, expressed in yen (¥). The term C_{ng} corresponds to the cost of natural gas, measured in yen per cubic meter (¥/m³). Additionally, Q_{ng} signifies the lower calorific value of natural gas, representing its energy content.

B. Battery Storage Model

During a specified time period Δt, the battery maintains a consistent rate of charging and discharging power. The change in the battery's State of Charge (SOC) is calculated using Equation (3). To ensure the battery is neither overcharged nor excessively discharged, the SOC must adhere to predefined upper and lower boundaries, as outlined in Equation (4).Since the battery cannot simultaneously charge and discharge, the charging and discharging operations must satisfy Equation(5).The battery energy state must satisfy the constraint of equality at the beginning and end of the scheduling period, as expressed in Equation(6).To maintain battery life, maximum charging and discharging power limits are established. This study establishes the maximum power output of the battery at 20% of its nominal capacity, as governed by the constraints specified in Equation (7).

$$S_t = S_0 + \frac{\sum_{t=1}^{T} P_{cha,t} X_t \Delta t - \sum_{t=1}^{T} P_{dis,t} Y_t \Delta t}{E_b} \quad (3)$$

$$S_{min} < S_t < S_{max} \quad (4)$$

$$X_t \cdot Y_t = 0 \quad (5)$$

$$S_0 = S_T \quad (6)$$

$$\begin{cases} 0 \leqslant P_{cha,t} \leqslant 0.2 E_b X_t \\ 0 \leqslant P_{dis,t} \leqslant 0.2 E_b Y_t \end{cases} \quad (7)$$

The parameters S_{max} and S_{min} indicate the upper and lower boundaries, respectively, of the battery's State of Charge (SOC) during time interval t. When the SOC attains its maximum threshold, the battery halts charging; conversely, when it reaches its minimum threshold, discharging is discontinued. The initial SOC of the battery is represented by S_0. The variables $P_{cha,t}$ and $P_{dis,t}$ denote the power associated with charging and discharging processes, respectively, within time period t. The binary variables X_t and Y_t, each taking values in $\{0,1\}$, signify the battery's charging and discharging states, respectively. The term Δt represents the duration of each time interval, while T indicates the total number of time periods considered. Finally, E_b corresponds to the total capacity of the battery.

III. OBJECTIVE FUNCTION

Microgrid dispatch aims to achieve coordinated optimal operation across three levels: the microgrid, distribution network, and user side. To satisfy the electricity demands of users, the power output from various distributed energy sources is strategically allocated to formulate an objective function aimed at reducing the operational expenses of the microgrid. The objective function F(t) encapsulates the total expenditure, which includes costs associated with fuel consumption and equipment maintenance, expenses for environmental pollution mitigation, and costs related to grid interaction. The objective function is formulated as:

$$F(t) = min \left\{ \begin{array}{l} \sum_{i=1}^{N} (C_i(P_i) + K_i \cdot P_i \cdot \Delta t) \\ + \sum_{i=1}^{N} \sum_{j=1}^{M} \alpha_j \cdot E_{i,j} \cdot P_i(t) \cdot \Delta t \\ + \sum_{t=1}^{N} c_{grid} |P_{grid}(t)| \end{array} \right\} \quad (10)$$

979-8-3315-2490-6/25 $31.00 © 2025 IEEE

$$c_{grid} = \begin{cases} c_{buy}(t), P_{grid}(t) \geqslant 0 \\ c_{sell}(t), P_{grid}(t) < 0 \end{cases} \quad (11)$$

The term $C_i(P_i)$ indicates the cost associated with power generation from microgrid source i, calculated based on the specific output characteristics of each energy source. The parameter K_i represents the coefficient for operation and maintenance expenses, which differs across various microgrid energy sources due to their distinct operational requirements, which varies among different microgrid sources; P_i is the output power of microgrid source i; α_j represents the conversion coefficient for pollutants; $E_{i,j}$ denotes the emission quantity of pollutant type j per unit power output from microgrid source i; The variable $P_{grid}(t)$ denotes the power exchanged between the microgrid and the main electrical grid during a specific time interval t. The parameter c_{grid} represents the cost associated with energy transactions between the microgrid and the main grid within the same time period t. When the microgrid procures electricity from the grid or supplies electricity to it, the transaction is governed by the time-of-use purchase price, $c_{buy}(t)$, and the time-of-use selling price, $c_{sell}(t)$.

IV. CONSTRAINTS

A. Power Balance Constraint

$$P_{load}(t) = P_{WT}(t) + P_{PV}(t) + P_{MT}(t)$$
$$+P_{BT}(t) + P_{grid}(t) \quad (12)$$

At time t, the microgrid's total load demand is denoted as $P_{load}(t)$. Renewable generation is represented by $P_{WT}(t)$ and $P_{PV}(t)$, corresponding to wind and solar power outputs, respectively. $P_{MT}(t)$ and $P_{BT}(t)$ refer to the power from the micro gas turbine and battery system. The variable $P_{grid}(t)$ indicates the power exchanged with the main grid through the tie-line. These variables form the basis for analyzing energy flow and optimizing microgrid operations.

B. Others Constraint

The main constraints involved are the power output constraints for battery turbines (BT), micro-turbines (MT), and tie-lines.

$$P_{\min} \leqslant P_{mg}(t) \leqslant P_{\max} \quad (13)$$

In the equation, $P_{\max}$ 、 $P_{\min}$ represent the upper and lower bounds of power output for battery turbines, micro-turbines, and tie-lines in the microgrid, (W).

V. EXPERIMENTAL ANALYSIS

The optimization methodology employed is a population-based metaheuristic inspired by collective behavior patterns observed in avian swarms during food-seeking activities. It is characterized by ease of implementation, rapid convergence, and minimal parameter tuning requirements. The algorithm process is illustrated in Figure 2.

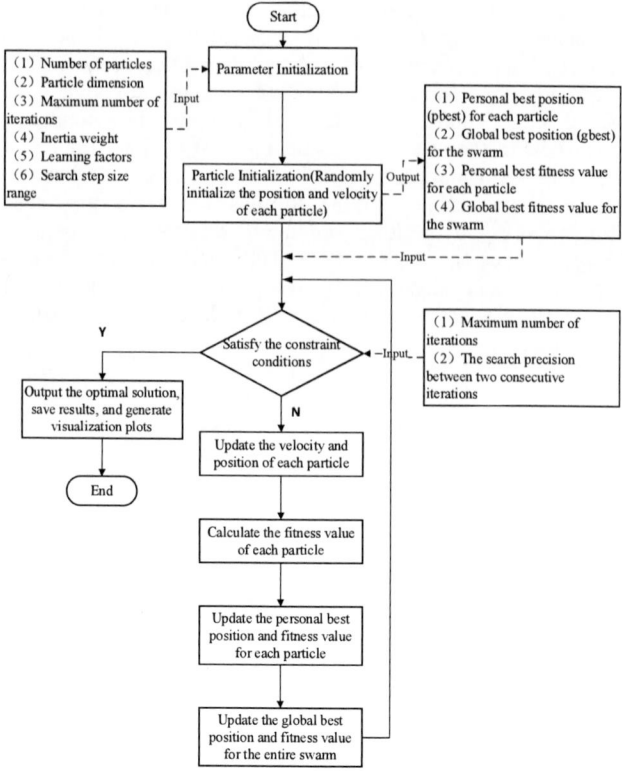

Figure 2. PSO Process

This study establishes a typical grid-connected microgrid system comprising PV, WT, MT, and BT components. The operational parameters for each DG unit in the microgrid are referenced from literature [7]. For assessing the effectiveness of the developed integrated scheduling approach for electrical networks incorporating operational flexibility limitations, Simulations were performed in MATLAB R2023b on a system equipped with an Intel i5-12600KF processor (3.7 GHz) and 32 GB RAM, providing sufficient computational resources for efficient model execution.

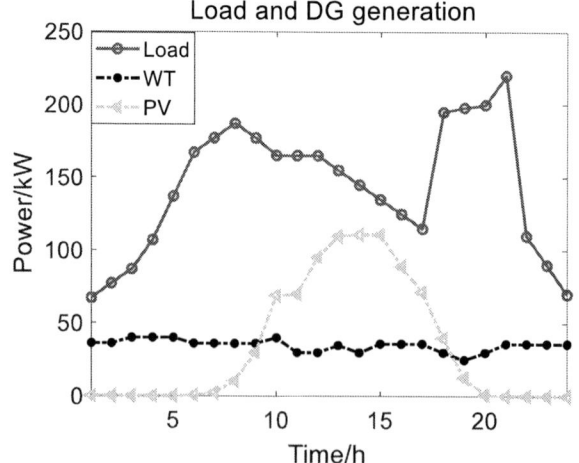

Figure 3. Load and DG Generation

979-8-3315-2490-6/25 $31.00 © 2025 IEEE

Photovoltaic and wind turbine units, being clean energy sources, incur negligible fuel costs and produce no pollutant emissions; therefore, only operation and maintenance costs need to be considered. The forecasted profiles for load, wind power, and photovoltaic output are presented in Figure 3.

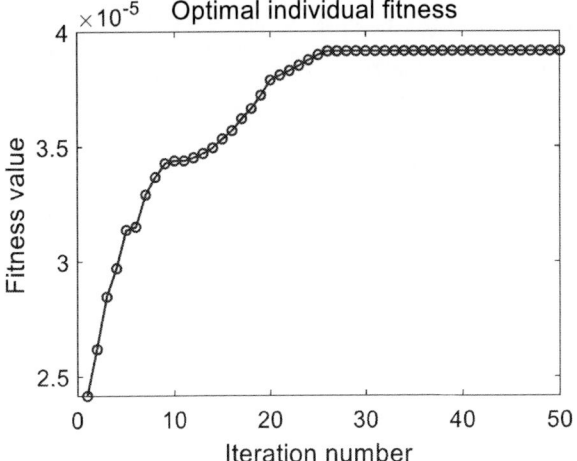

Figure 4. Optimal Individual Fitness

The particle swarm optimization (PSO) algorithm was configured with 10 particles and 50 iterations. The acceleration coefficients were set to $c_1 = 0.8$ and $c_2 = 0.9$, with an inertia weight of $w = 0.8$. The algorithm optimizes both environmental protection costs and microgrid operation and maintenance costs. As shown in Figure 4, the fitness function converges after approximately 16 iterations.

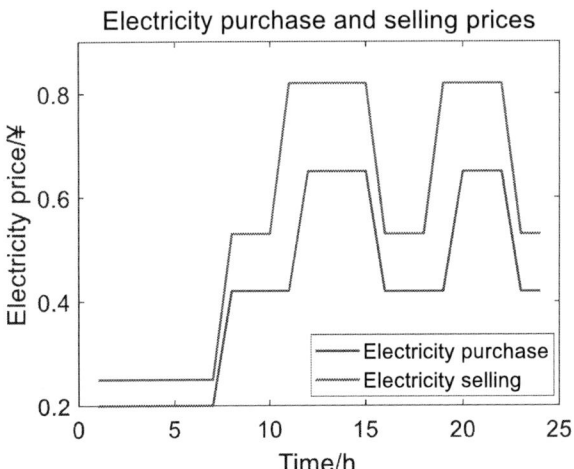

Figure 5. Electricity purchase and selling prices

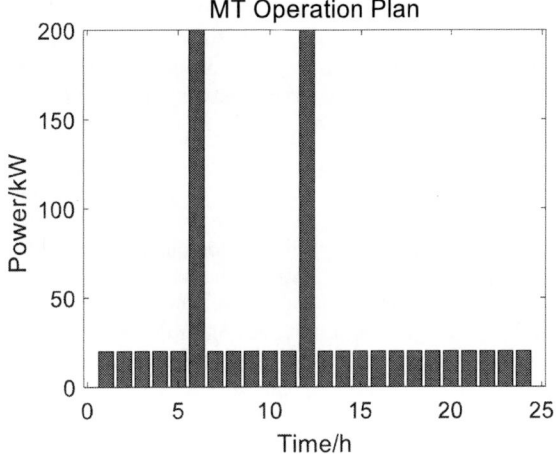

Figure 6. MT Operation Plan

Figure 7. BT Operation Plan

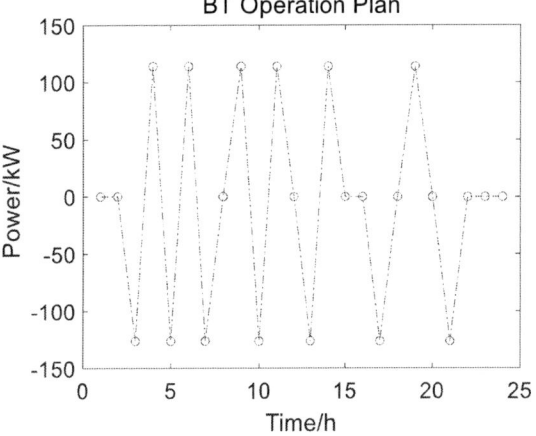

Figure 8. grid Operation Plan

In this case study, photovoltaic and wind generation are fully utilized within the microgrid. Economic optimization is achieved through the coordinated scheduling of micro gas turbine output, battery charging/discharging, and main grid interactions. Under the Electricity purchase and selling prices as show in Figure 5、figure 6、figure 7 and figure 8 shows the optimal power dispatch results for each unit under the grid-connected mode, highlighting coordinated system operation.

The obtained scheduling solution demonstrates that battery energy storage effectively addresses power demand during peak load periods through auxiliary regulation, thereby enhancing grid stability and security. The dispatch strategy schedules battery charging during low-price periods and discharging during high-price intervals, improving both cost-efficiency and environmental performance. Due to the intermittency of solar and wind power, the microgrid relies on either the utility grid or distributed generation to meet residual load demand. When the comprehensive benefits of gas turbine generation exceed those of purchasing power from the main grid, gas turbine generation is prioritized. When the microgrid's internal generation exceeds load demand, surplus electricity is supplied back to the main grid.

VI. Conclusion

This paper proposes an optimization dispatch model for microgrid systems comprising wind turbines, photovoltaic generation, micro gas turbines, and energy storage, with the objective of minimizing total operational costs. The model is solved using particle swarm optimization (PSO) algorithm. The key findings are summarized as follows:(1) The PSO algorithm demonstrates efficient convergence within only 16 iterations using a population size of 10 particles, indicating robust optimization capability; (2) The proposed dispatch model effectively addresses the challenges posed by renewable energy uncertainty. Furthermore, the scheduling strategy successfully implements time-of-use electricity pricing optimization by charging batteries during off-peak periods and discharging during peak hours, thereby significantly reducing operational costs.

However, this study has certain limitations. The research considers only a single microgrid configuration without addressing multi-microgrid interactions or demand response mechanisms. Future research could extend the framework to encompass multi-microgrid interactive systems. Additionally, the model assumes perfect forecasting of renewable energy generation and load demand, which may not accurately reflect real-world uncertainties. Subsequent studies could incorporate probabilistic forecasting methods to better handle the stochastic nature of renewable energy sources.

References

[1] Yushuang Wen, Ran Ding, Yaquan Liang, Jiaqi Chen, "Multi-microgrid shared energy storage operation optimization strategy based on a mixed game theory, "Electric Power Systems Research,15 (2), pp. 803 - 818,2025.

[2] Peter Anuoluwapo Gbadega, Yanxia Sun, Olufunke Abolaji Balogun, "Optimized energy management in Grid-Connected microgrids leveraging K-means clustering algorithm and Artificial Neural network models, "Energy Conversion and Management, Volume 336,2025.

[3] W. Chen, Z. Wang, Q. Ge, H. Dong and G. -P. Liu, "Quantized Distributed Economic Dispatch for Microgrids: Paillier Encryption – Decryption Scheme," IEEE Transactions on Industrial Informatics, vol. 20, no. 4, pp. 6552-6562, April 2024.

[4] Y. Wang, M. Xiao, Y. You and H. V. Poor, "Optimized Energy Dispatch for Microgrids With Distributed Reinforcement Learning," IEEE Transactions on Smart Grid, vol. 15, no. 3, pp. 2946-2956, May 2024.

[5] H. Li, Q. Wu, L. Yang, H. Zhang and S. Jiang, "Distributionally Robust Negative-Emission Optimal Energy Scheduling for Off-Grid Integrated Electricity-Heat Microgrid,"IEEE Transactions on Sustainable Energy, vol. 15, no. 2, pp. 803-818, April 2024.

[6] Y. Zou, Y. Xu and C. Zhang, "A Risk-Averse Adaptive Stochastic Optimization Method for Transactive Energy Management of a Multi-Energy Microgrid," IEEE Transactions on Sustainable Energy, vol. 14, no. 3, pp. 1599-1611, July 2023.

[7] Sun J X, Wang P Y, Hu X Y, et al. Research on improved genetic algorithm for micro-grid grid-connected multi-objective optimization scheduling[J]. Northeast Electric Power Technology, 2018, 39(9): 32-35, 43.

2025 International Conference on Advanced Energy Systems and Power Electronics (AESPE 2025)

A Wavelet-Enhanced Diagnostic Framework for Internal Short Circuit Detection in Lithium-Ion Batteries Using EIS and XGBoost

1st Jilong Song
School of Electrical
Engineeringy
Qingdao University
Qingdao, China
songjilong_x@163.com

2nd Jinshuo Fu
School of Drake United
Qingdao University
Qingdao, China
15054372367@163.com

3rd Songtao Che
School of Marine
Engineering
Dalian Maritime University
Dalian, China
18669706996@163.com

4th Hongsi Shi
Qingdao Rectifier
Manufacturing Co.
Qingdao, China
shs313@126.com

5th Kai Wang
School of Electrical
Engineering
Qingdao University
Qingdao, China
Corresponding Author:
wkwj888@163.com

Abstract—With the widespread application of lithium-ion batteries in electric vehicles and energy storage systems, safety issues have become increasingly prominent, particularly internal short circuit (ISC) faults, which can easily lead to thermal runaway and serious accidents. This paper proposes an ISC fault diagnosis method based on electrochemical impedance spectroscopy (EIS) data. The method first uses continuous wavelet transform (CWT) to extract multi-scale frequency domain energy features and combines battery state of charge (SOC) and state of health (SOH) to construct multi-dimensional feature vectors. Subsequently, an XGBoost classifier is employed for ISC identification, and Bayesian hyperparameter optimisation is conducted using the Optuna framework. The proposed method was validated using impedance data from 600 NCM811 batteries under simulated short-circuit and normal conditions at different SOH and SOC conditions. Experimental results on 600 impedance samples achieve an accuracy of 94% and recall of 95%, demonstrating the method's robustness across varying SOH and SOC conditions. The proposed method exhibits excellent classification performance, providing a reliable data-driven solution for early fault warning of lithium-ion batteries.

Keywords—*EIS, ISC fault diagnosis, XGBoost, Parameter optimisation, Battery safety*

I. INTRODUCTION

With the rapid development of electric vehicles and energy storage systems, lithium-ion batteries have emerged as the most competitive electrochemical energy storage solution due to their high energy density, long cycle life, and low self-discharge rate[1]. However, battery thermal safety remains the core challenge hindering its further adoption, particularly the failure mode of internal short circuits, which are highly concealed and destructive[2]. These failures often originate from minor internal structural damage, leading to thermal runaway and resulting in severe safety incidents such as fires or explosions[3].

According to statistics, Internal short circuit (ISC) is one of the primary causes of battery safety incidents[4, 5]. In recent years, electrochemical impedance spectroscopy (EIS) has gained attention due to its ability to provide frequency-domain responses of multi-physical processes within batteries[6, 7]. By analysing impedance characteristics at different frequencies, changes in battery internal structure and electrochemical processes can be identified, enabling early warning of faults[8, 9].

However, EIS spectra are high-dimensional, non-linear, and easily influenced by state of health (SOH) and state of charge (SOC) changes, making them less effective for direct fault identification[10, 11]. Therefore, this paper proposes an ISC detection scheme that integrates continuous wavelet transform (CWT) multi-scale energy spectrum analysis with the XGBoost method. By integrating wavelet energy with SOH/SOC features to construct more robust feature vectors, followed by XGBoost for classification tasks and Optuna for hyperparameter tuning, this scheme achieves high-precision detection of ISC faults.

To systematically elucidate the principles and effectiveness of the proposed method, the remainder of this paper is organised as follows: Section II describes the proposed diagnostic method, including CWT energy spectrum extraction, wavelet feature fusion, XGBoost classification model construction, and Optuna hyperparameter optimisation mechanism; Section III introduces the dataset, detailing the selection of experimental batteries, EIS data acquisition scheme, and design of simulated short-circuit conditions; Section IV presents the model detection results and analysis, verifying the ISC identification performance and generalisation ability of the method; Section V summarises the entire paper and discusses the engineering application prospects and future extension directions of this study.

II. METHODS

This study proposes a lithium-ion battery ISC detection method based on impedance spectroscopy. The overall framework consists of four core steps: CWT energy spectrum extraction, feature fusion, XGBoost construction, and hyperparameter optimisation.

A. CWT

The CWT is a multiresolution analysis method that maps the original signal onto a scale–translation plane, thereby revealing its local frequency and time domain characteristics simultaneously. Performing CWT on the impedance amplitude sequence $|Z(f)|$ yields the wavelet coefficients：

$$W(a,b) = \int_{-\infty}^{\infty} |Z(f)| \overline{\psi_{a,b}(f)} df \qquad (1)$$

Among them, $\psi_{a,b}(t) = \frac{1}{\sqrt{a}} \psi\left(\frac{t-b}{a}\right)$ is the wavelet function of scale a and time shift b; a is inversely proportional to pseudo frequency $f_p = \frac{f_c f_s}{a}$, f_c refers to the peak of the equivalent frequency response of the selected mother wavelet at unit scale, and f_s refers to the highest frequency of the original EIS measurement data. The larger the scale, the lower the corresponding frequency.

The Mexican Hat, which is the second derivative of a Gaussian function, can sensitively capture local extrema and peak-like sudden changes in signals:

$$\psi(t) = \frac{2}{\sqrt{3}\sigma \pi^{1/4}}\Big(1 - \frac{t^2}{\sigma^2}\Big)e^{-t^2/(2\sigma^2)} \tag{2}$$

Then, the wavelet energy at each scale is defined as:

$$E(a) = \int_{-\infty}^{\infty} |W(a,b)|^2 \, \mathrm{d}b \tag{3}$$

That is, integrate the square of the coefficients for all translation positions on the same scale. Here, $E(a)$ denotes the characteristic energy at scale a. After converting the scale to pseudo-frequency, plot the relationship curve between E and f_p to visually compare the energy differences between the healthy baseline and short-circuit samples across all frequency bands: internal short circuits introduce additional fluctuations at low frequencies, causing significant shifts in energy in that frequency band; Mexican hat wavelets are particularly sensitive to second-order structures, further amplifying these fault characteristics.

B. Add SOH and SOC features

In order to further improve the adaptability of the model under different health states and charge states, the two parameters SOH and SOC corresponding to each sample are concatenated as scalar features to the energy spectrum features to form the final input vector:

$$x = [\log(1 + E_1), ..., \log(1 + E_{64}), SOH, SOC] \tag{4}$$

Among them: E_i: wavelet energy corresponding to the i-th scale; SOH: value range is 0–1; SOC: value range is 0–1.

C. XGBoost binary classification model

XGBoost is based on gradient boosting and completes classification tasks by integrating multiple decision trees[12]. Its optimisation objective consists of a loss function and a regularisation term:

$$L^{(t)} = \sum_{i=1}^{n} \ell\big(y_i, y_i^{(t-1)} + f_t(x_i)\big) + \Omega(f_t) \tag{5}$$

Among them: y_i: the true label of sample i, where 1 indicates healthy and 0 indicates short circuit; $\hat{y}_i^{(t-1)}$ is the probability value output by the previous round model; $f_t(x_i)$ is the prediction output of the current weak learner; ℓ is the logarithmic loss function:

$$\ell(y, \hat{y}) = -y\log(\hat{y}) - (1-y)\log(1-\hat{y}) \tag{6}$$

$\Omega(f) = \gamma T + \frac{1}{2}\lambda \sum_{j=1}^{T} w_j^2$ is a regularisation term. In the regular term, T is the number of leaf nodes in the current tree; w_j is the output weight of the j-th leaf node; γ, λ is the regularisation hyperparameter. The model outputs a probability

of $\hat{y} \in [0,1]$, and the initial classification threshold is 0.5. To further improve detection performance, the Youden index is introduced: $J(\theta) = \mathrm{TPR}(\theta) - \mathrm{FPR}(\theta)$, where TPR is the true positive rate; FPR is the false positive rate; θ is the decision threshold.

D. Optuna Hyperparameter Optimization Principles

To improve model generalisation performance and reduce manual parameter tuning costs, this paper uses the Optuna framework to perform Bayesian optimisation of XGBoost hyperparameters, with the optimisation objective being to maximise the mean cross-validation AUC:

$$\text{maximize } \mathbb{E}_{\mathrm{CV}}[\mathrm{AUC}(\mathrm{XGB}(\theta))] \tag{7}$$

Among them: θ is the XGBoost parameter vector; $\mathrm{XGB}(\theta)$ is the model trainer under the current parameters; $\mathbb{E}_{\mathrm{CV}}$ is the expectation on 5-fold GroupKFold.

The TPE sampler constructs two sets of parameter distributions:

$l(x)$: Kernel density estimation of low-performance test samples; $g(x)$: Kernel density estimation of high-performance test samples.

Each round samples from the region with the maximum ratio of $\frac{g(x)}{l(x)}$. In addition, Optuna terminates inefficient trials through an early-stopping pruning mechanism, accelerating the optimisation process and improving search quality.

III. DATA SET

Data was collected to measure the EIS of NCM811 batteries at five different State of Health (SOH) levels (0.8, 0.85, 0.9, 0.95, and 1) and ten different State of Charge (SOC) levels for each SOH[13]. The EIS measurements were tested on an electrochemical workstation (Autolab, AUT86219). Ten SOC points were selected, ranging from 10% to 100% (every 10%), for EIS testing. At each SOC point, 12 sets of impedance data were collected (including 6 normal sets and 6 simulated short-circuit sets), corresponding to different equivalent short-circuit resistances. Internal short circuits are simulated by connecting external resistors with different resistance values in parallel across the battery terminals. The specific resistance values include: 200Ω, 100Ω, 50Ω, 30Ω, 20Ω, and 10Ω. The details of impedance data collection at each SOC point are shown in Table I.

TABLE I. THE DETAILED TEST INFORMATION OF 12 EIS MEASUREMENTS IN EACH SOC.

Test number	Test type	Equivalent ISC resistance	Test number	Test type	Equivalent ISC resistance
1	Normal test	-	2	ISC	200Ω
3	Normal test	-	4	ISC	100Ω
5	Normal test	-	6	ISC	50Ω

7	Normal test	-	8	ISC	30Ω
9	Normal test	-	10	ISC	20Ω
11	Normal test	-	12	ISC	10Ω

EIS testing was conducted in the frequency range of 0.01 Hz to 1 MHz, with each impedance curve containing 81 frequency points corresponding to different physical-chemical process time scales.

IV. RESULTS AND DISCUSSION

Based on the proposed CWT + Optuna–XGBoost method, the ISC of batteries is detected. The algorithm flowchart is shown in Fig.1. The entire diagnostic process begins with precise EIS measurements conducted under controlled conditions, followed by feature processing and machine learning classification. The selection of hyperparameters is based on multiple preliminary experiments, and Optuna performs Bayesian optimisation of XGBoost hyperparameters. Each stage is designed to capture and utilise significant features to achieve accurate ISC detection. Among the total 600 impedance samples, we divided the data into training and testing sets in an 80%/20% ratio. The testing set contains 120 data points with balanced labels, drawn from batteries with different SOH (1.00–0.80), different SOC (10%–100%), and two categories (normal and short-circuited). Data processing and model construction were implemented using Python 3.12 and Windows 11. Computations were performed on Intel Core i5-13490F CPU, 32 GB RAM, and an NVIDIA GeForce RTX 4060 graphics processing unit.

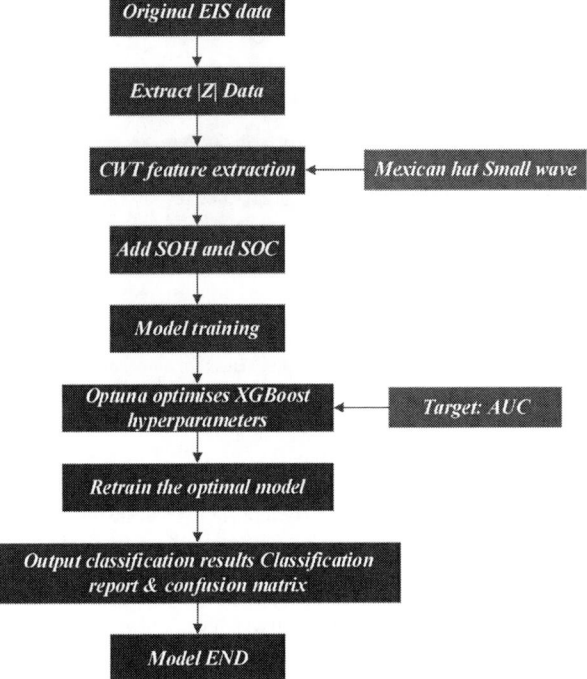

Fig. 1. CWT + Optuna–XGBoost ISC detection algorithm flowchart

As shown in Fig.2, taking the impedance curve obtained under the conditions of SOH = 1 and SOC = 10% as an example, there are slight differences between the impedance curves of normal samples and ISC samples. The occurrence of ISC introduces a low-impedance parallel branch into the battery equivalent circuit, equivalent to adding a parallel resistor. This bypass allows low-frequency signals to bypass the normal charge transfer process route, resulting in a significant change in the 'tail' of the polarisation diffusion in the low-frequency segment of the impedance spectrum. The sudden change in impedance causes a noticeable difference in the impedance curve. However, this change is not significant numerically, making classification based on manual or threshold methods impractical.

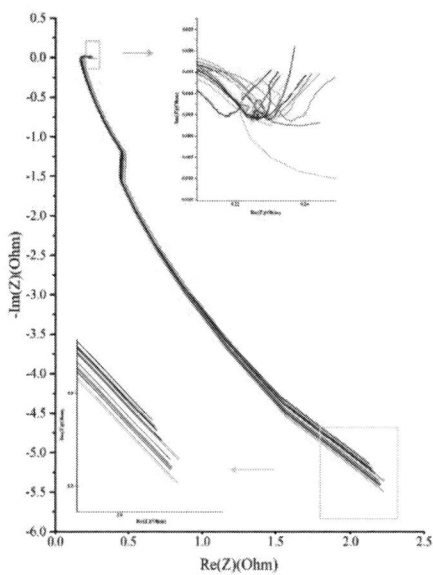

Fig. 2. Data group battery impedance curve when SOH = 1/SOC = 10%

As a multi-scale decomposition tool, CWT can decompose the impedance response into energy distributions at different scales. With its excellent time-frequency localisation characteristics, CWT can highlight local mutations or morphological changes. Low-frequency impedance mutations caused by internal short circuits will manifest as concentrated fluctuations in energy at the corresponding scale. This method ensures that local anomalies in the signal domain are presented as characteristic concentrations in the wavelet energy domain, thereby effectively amplifying fault features that were previously difficult to detect, making them easier to detect automatically.

Fig.3 shows the *E-f* curves of each battery in the same battery pack after CWT. The differences between faulty batteries and normal batteries after CWT increase, exhibiting more distinct fault characteristics. At low frequencies, the influence of noise intensifies, making the ISC characteristics more pronounced[14].

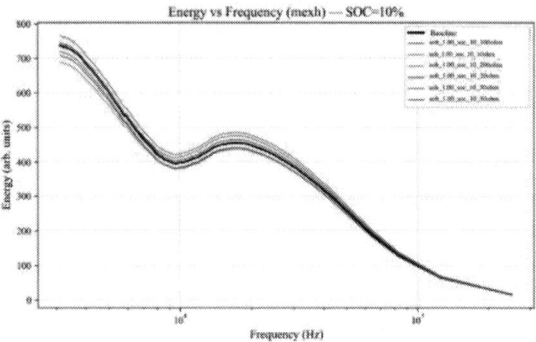

Fig. 3. When SOH = 1/SOC = 10%, the CWT decomposition E-f curve of the battery data set.

By amplifying fault features through CWT[15], introducing SOH and SOC as scalar features, and then using the XGBoost model for diagnostic testing, the confusion matrix of the classification results obtained after hyperparameter optimization using Optuna on the test set is shown in the Table II:

TABLE II. CONFUSION MATRIX

	Predicted Normal	Predicted Fault
True Normal	60	0
True Fault	3	57

Calculate Accuracy according to the following formula:

$$Accuracy = \frac{TP+TN}{TP+TN+FP+FN} = \frac{60+57}{120} = 0.94 \quad (8)$$

The results indicate that the model can accurately identify most short-circuit samples while ensuring zero false positives. Only three ISC samples were incorrectly classified as normal. As shown in Fig.4(b), the accuracy, recall and Precision rates reached 0.94, 0.95 and 1, respectively, indicating that the model possesses excellent state generalization capability. As shown in Fig.4(c), the proposed CWT-XGBoost-based ISC diagnostic method maintains high classification accuracy under various SOH and SOC combinations. By capturing local, scale-specific energy changes in the impedance spectrum, this method can reveal subtle fault features that are often overlooked in traditional frequency domains. Even under low SOH or extreme SOC conditions, the model maintains good performance, achieving ideal classification accuracy in most scenarios. This highlights the model's robustness and generalisation capability when faced with diverse operating conditions. The above findings indicate that wavelet-based multi-scale analysis provides a novel framework for battery ISC fault diagnosis, enabling precise identification of ISC faults by fully utilising EIS curves.

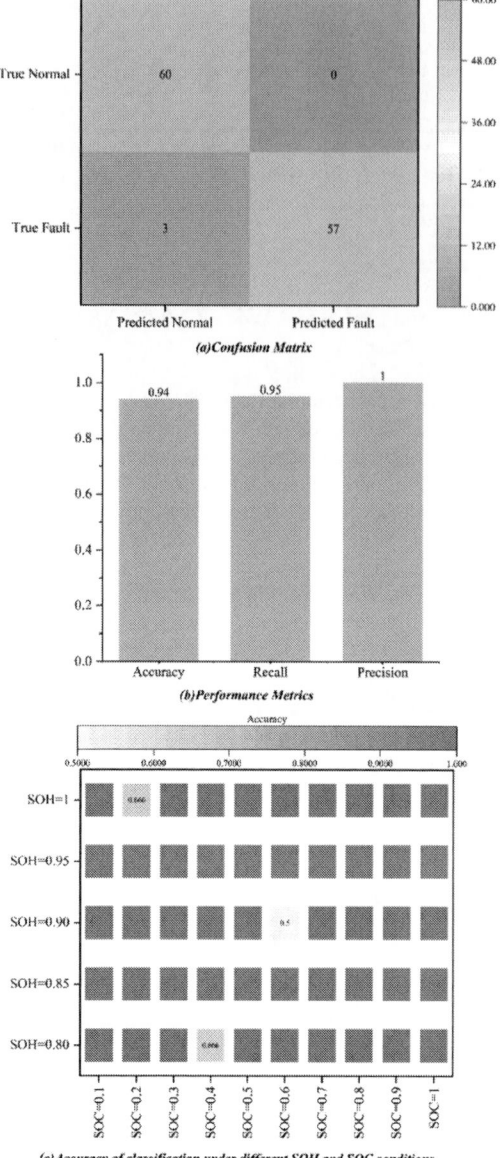

Fig. 4. Model test results. (a). Confusion matrix for model classification (b). Model performance metrics: accuracy, precision and recall (c). Accuracy of classification under different SOH and SOC conditions in a random partitioning scenario

Although the proposed wavelet-based diagnostic framework demonstrates high accuracy in detecting ISC in the dataset tested in this paper. However, the dataset used in this study is relatively limited, and the limited dataset imposes constraints on fault severity classification, validation of generalizability across broader temperature and operating ranges, and integration with other machine learning models. Therefore, future research should focus on expanding the dataset scale and improving data quality, testing fault data under various conditions such as different temperatures and ageing to further enhance model performance. Additionally, optimising the model for deployment within the BMS is an important future research direction.

979-8-3315-2490-6/25 $31.00 © 2025 IEEE

V. Conclusion

This paper proposes a battery internal short-circuit fault diagnosis method based on the CWT and Optuna–XGBoost framework. By performing a wavelet transform on EIS data to extract energy spectra, and combining these with battery SOH and SOC features, a multi-dimensional input feature set is constructed. XGBoost is then used for classification, followed by Optuna for model optimization. The method achieves good recognition performance on 600 sample data points, validating its feasibility and general applicability.

Experimental results demonstrate that the method maintains high robustness under conditions of declining SOH and deteriorating battery performance, while also exhibiting good generalization capabilities. Future work can be extended to multi-type fault identification and online detection deployment to establish a more comprehensive battery safety management system.

References

[1] J. H. Ren, J. K. Ma, H. H. Wang, T. Yu, and K. Wang, "A comprehensive review on research methods for lithium-ion battery of state of health estimation and end of life prediction: Methods, properties, and prospects," Protection Control of Modern Power Systems, vol. 10, no. 3, pp. 146 - 165, May 2025.

[2] K. Zhao, Z. Y. Yan, and K. Wang,"Research Progress of Fiber Optic Sensing Technology in Battery Charge State/Health Status Monitoring," Power Generation Technology, pp. 1-12, Mar 2025.

[3] Y. Y. Ma, X. Y. Wang, H. Yuan, G. F. Chang, J. G. Zhu, H. F. Dai, and X. Z. Wei, "Review of electrochemical impedance spectroscopy in fault diagnosis for proton exchange membrane fuel cells," Renewable & Sustainable Energy Reviews, vol. 211, Apr 2025, Art. no. 115226.

[4] X. Y. Chen, Y. F. Feng, J. N. Shen, and Y. J. He, "Quantitative Diagnosis of Early Internal Short Circuit for Lithium-Ion Batteries Based on Multirate Discharge," Ieee Transactions on Instrumentation and Measurement, vol. 74, 2025, Art. no. 3521012.

[5] Z. Wang, Q. J. Zhao, S. J. Wang, Y. C. Song, B. B. Shi, and J. J. He, "Aging and post-aging thermal safety of lithium-ion batteries under complex operating conditions: A comprehensive review," Journal of Power Sources, vol. 623, Dec 2024, Art. no. 235453.

[6] K. Ariyoshi, "Electrochemical Impedance Analysis of Lithium Insertion Electrodes Using Symmetric Cells," Electrochemistry, vol. 92, no. 3, 2024, Art. no. 037007.

[7] L. A. Santa-Cruz, F. C. Tavares, L. F. Loguercio, C. I. L. dos Santos, R. A. Galvao, O. A. L. Alves, M. Z. Oliveira, R. M. Torresi, and G. Machado, "Electrochemical impedance spectroscopy: from breakthroughs to functional utility in supercapacitors and batteries - a comprehensive assessment," Physical Chemistry Chemical Physics, vol. 26, no. 40, pp. 25748-25761, Oct 2024.

[8] J. H. Wu, W. Q. Bai, L. P. Zhang, X. Y. Zhang, H. J. Lin, H. D. Dai, J. J. Liu, F. Zhang, and Y. X. Yang, "Design of a portable electrochemical impedance spectroscopy measurement system based on AD5941 for lithium-ion batteries," Journal of Energy Storage, vol. 84, Apr 2024, Art. no. 110856.

[9] W. C. Zhang, Z. R. Long, L. Y. Zhuang, H. C. He, Y. Xie, and J. J. Zhou, "Early warning and severity classification of lithium-ion battery internal short circuits using cosine transform and image coding," Measurement, vol. 246, Mar 2025, Art. no. 116663.

[10] J. Xie and Y. C. Lu, "Designing Nonflammable Liquid Electrolytes for Safe Li-Ion Batteries," Advanced Materials, vol. 37, no. 2, Jan 2025.

[11] K. Panchal, K. Bhakar, K. S. Sharma, D. Kumar, and S. Prasad, "Review on electrochemical impedance spectroscopy: a technique applied to hollow structured materials for supercapacitor and sensing applications," Applied Spectroscopy Reviews, vol. 60, no. 1, pp. 30-55, Jan 2025.

[12] Z. Y. Yan, K. Zhao, and K. Wang,"A Review of Novel Power System Optimization Strategies Based on Reinforcement Learning," Power Generation Technology, pp. 1-13, May 2025.

[13] B. H. Cui, H. Wang, R. L. Li, L. Z. Xiang, J. N. Du, H. A. Zhao, S. Li, X. Y. Zhao, G. P. Yin, X. Q. Cheng, Y. L. Ma, H. Huo, P. J. Zuo, and C. Y. Du, "Internal short circuit early detection of lithium-ion batteries from impedance spectroscopy using deep learning," Journal of Power Sources, vol. 563, Apr 2023, Art. no. 232824.

[14] X. T. Zhu, Z. R. Wang, Z. M. Fang, Z. Xu, B. Q. Luo, H. X. Yang, K. Wang, and B. Guo, "Flame-retardant polymer electrolytes enhancing the safety of lithium batteries," Journal of Energy Storage, vol. 108, Feb 2025, Art. no. 115080.

[15] H. Jiang, and K. Wang, "A Review of Multi-Source Fusion Real-Time Localization and Mapping Methods for Unmanned Patrols in Substations," Guangdong Electric Power, pp. 1-16, May 2025.

Multi-distributed energy synergistic regulation method considering demand response

Chaoyang Zhi[1]

[1]Shenyang Ligong University, Shenyang, 110000, China

*Corresponding author's e-mail: 17839691576@163.com

Abstract-The convergence of Industrial Internet of Things and smart manufacturing has driven the evolution of energy systems in the direction of intelligence and distribution. Among them, the large-scale access of distributed energy resources (DER) leads to volatility and load peak-to-valley difference, which exacerbates the difficulty of grid regulation. Aiming at the microgrid containing electric vehicles (EVs) and energy storage, this paper thoroughly studies the adjustable load characteristics of Demand Response (DR), and proposes a cooperative regulation method with multiple distributed energy resources. First, a stochastic scenario of EV charging demand is generated based on Monte Carlo simulation to consider the disordered charging behavior when EVs are connected and evaluate its impact on the peak-to-valley load difference. Second, combining the regulation potentials of Shiftable Load and Interruptible Load, a multi-objective mixed integer planning (MILP) model for distributed energy aggregation systems considering demand response is constructed, and a nonlinear quadratic penalty term is proposed to suppress load fluctuations. Solved by CPLEX tool, the results show that the method can reduce the peak-to-valley difference of typical bimodal load background by about 25.4%, reduce the operating cost significantly, coordinate demand response, electric vehicles and energy storage effectively, improve the system flexibility, and provide a reference for distributed energy aggregation control.

Keywords—Time-of-Use Tariff; Demand Response; Distributed Energy Resources; Electric Vehicle

I. Introduction

Large-scale access to distributed energy sources is reshaping the traditional power system operation model[1]. The volatility of renewable energy sources such as wind power and photovoltaic (PV) challenges the stability and economy of microgrids under high penetration conditions[2]. The popularity of electric vehicles further exacerbates the load peak-to-valley difference and increases the scheduling difficulty[3]. The development of technologies such as industrial internet of things and information physical systems provides new ideas for distributed energy sensing and cooperative control. Achieving source-load-storage cooperative optimization to enhance the flexibility of system operation and renewable energy consumption capacity has become a key issue in current research.

Demand Response (DR), as an important means to enhance load flexibility[4], is mainly divided into price-based and incentive-based, while energy storage has the characteristics of flexible control of energy input and output, and can respond to scheduling commands on different time scales[5], and the literature[6] integrated demand response shaves peaks and fills valleys of loads, which enhances energy utilization efficiency. Literature[7] proposes an optimal dispatch model for integrated energy systems that incorporates a combined electricity and heat demand response model, and verifies the effectiveness of the wind power integrated demand response strategy through simulation. Literature[8] formulates response priorities for demand-side adjustable resources, and calls resources according to the priorities during demand response. Literature[9] designed a real-time demand response scheme incorporating blockchain technology. Most of the above studies treat demand-side adjustable resources as deterministic resources and focus on a single time scale optimization. The coupling characteristics of DR and EV charging scheduling, storage lifetime loss, and other factors have not been fully quantified. These limitations make it difficult for existing methods to adapt to the strong coupling and uncertainty of multi-timescale decision variables in high percentage renewable energy scenarios.

This paper proposes a coordinated regulation method for distributed energy resources considering demand response. The model integrates flexible loads, simulates EV charging via Monte Carlo, and introduces a penalty term to smooth power fluctuations and reduce costs. Simulation results confirm the method's effectiveness.

II. Distributed Energy Modeling

A. Diesel Engine Model

Diesel generators are flexible but inefficient and polluting, so they are mainly used as a backup power source. Diesel cost model is often expressed as a quadratic function or segmented linear function, the cost function as shown in equation (1)

$$C_{DG}(t) = a \cdot u_{DG}(t) + b \cdot P_{DG}(t) + c \cdot z_{DG}(t) \tag{1}$$

B. PV model

Photovoltaic power generation is characterized by modular structure, light weight flexibility and adaptability. Its predicted value is shown in equation (2)(3)

$$P_{pv}(t) = P_{pv,rated} \cdot f_{PV}(G, T_c) \tag{2}$$

$$f_{PV}(G, T_c) = \frac{G}{G_{STC}}[1 + \alpha(T_c - T_{STC})] \tag{3}$$

C. Wind power modeling

The size of WT output is closely related to wind speed and weather conditions. The relationship between the predicted WT output and wind speed is shown in equation (4).

$$P_{wt}(v) = \begin{cases} 0, & v < v_{in} \text{ or } v > v_{out} \\ P_{rated} \dfrac{v^3 - v_{in}^3}{v_{rated}^3 - v_{in}^3}, & v_{in} \le v < v_{rated} \\ P_{rated}, & v_{rated} \le v \le v_{out} \end{cases} \tag{4}$$

D. Adjustable Flexible Load Model

Flexible loads consist of transferable loads as well as curtailable loads. Flexible loads are required to satisfy the constraints of constant total load, power change constraints and interruption ratio constraints as shown in equations (5)(6)(7).

$$\sum_{t=1}^{24} L_{shift}(t) = \sum_{t=1}^{24} L_0(t) \tag{5}$$

$$-\alpha \cdot L_0(t) \le L_{shift}(t) - L_0(t) \le \alpha \cdot L_0(t) \tag{6}$$

$$0 \le L_{int}(t) \le \beta \cdot L_0(t) \tag{7}$$

E. Energy storage system

The power balance equation of the energy storage system is shown in equation (8)

$$SOC(t) = SOC(t-1) + \frac{P_{ch}(t)\eta_{ch} - P_{dis}(t)/\eta_{dis}}{E_{cap}} \Delta t \tag{8}$$

F. Electric vehicle charging load modeling

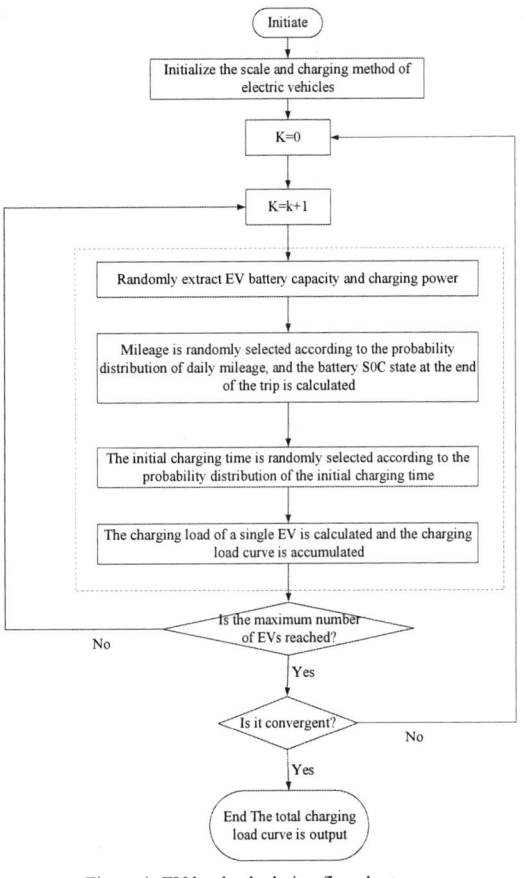

Figure 1. EV load calculation flowchart.

The EV charging load is large and concentrated, greatly exacerbating the peak-to-valley difference of the grid load, and the EV users' charging behavior is stochastic, which is simulated by Monte Carlo simulation. The calculation process is shown in Figure 1, and the probability density function of the stochastic distribution of EV single-vehicle daily mileage is shown in Equation (9)

$$f(x \mid \mu, \sigma) = \frac{1}{x\sigma\sqrt{2\pi}} \exp\left(-\frac{(\ln x - \mu)^2}{2\sigma^2}\right) \tag{9}$$

The Monte Carlo random sampling generates the single vehicle daily mileage for the single vehicle daily charging demand calculation formula as shown in equation (10)(11)

$$Q_{EV,i} = \frac{KM_i}{\eta_{EV}} \tag{10}$$

$$P_{EV}(t) = \sum_{i=1}^{N} P_{EV,i}(t) \tag{11}$$

III. MULTI-ENERGY SYNERGISTIC REGULATION METHOD CONSIDERING DEMAND RESPONSE

In this paper, we study the construction of an economic cost

979-8-3315-2490-6/25 $31.00 © 2025 IEEE

optimization model for a distributed energy aggregation system containing electric vehicle charging, demand response, and multiple types of distributed energy sources (wind power, photovoltaic, diesel, and energy storage).

A. Objective function

The time scale of the multi-collaborative scheduling model is 1 h, and the scheduling period is 24 h. The model takes the objective of minimizing the total operation cost, including the diesel operation cost, DR cost, wind abandonment penalty cost, and the interaction cost with the main grid. Among them, the diesel engine operation cost includes startup cost, fuel cost and operation and maintenance cost, and the objective function is shown in equations. (12)-(18)

$$F = \sum_{i}^{5} F_i + F_{smooth} \tag{12}$$

$$F_1 = \sum_{t=1}^{24} (P_{mgb}(t) \cdot x_b(t) - P_{mgs}(t) \cdot x_s(t)) \tag{13}$$

$$F_2 = \sum_{t=1}^{24} (a + k_{cp} \cdot P_{mt}(t) + s_{conv} \cdot y_{conv}(t)) \tag{14}$$

$$F_3 = \sum_{t=1}^{24} \sum_{m=1}^{3} pil(m,t) \cdot kil(m) \tag{15}$$

$$F_4 = cost_{ev} \cdot \frac{\sum_{t=1}^{24} EV_{charge}(t)}{\eta_{charge}} \tag{16}$$

$$F_5 = \sum_{t=1}^{24} -0.5 \cdot k_{shift}(t) \cdot (shiftload(t) - P_{load}(t)) \tag{17}$$

$$F_{smooth} = \alpha \cdot \sum_{t=1}^{24} (EV_{charge}(t))^2 \tag{18}$$

B. Constraints

The constraints include power balance constraints, diesel engine constraints, wind power output constraints, battery storage system constraints, DR constraints, and interactive power constraints.

The system power balance constraint is shown in equation (19).

$$P_{wt}(t) + P_{pv}(t) + P_{DG}(t) + P_{buy}(t) + P_{dis}(t) = P_{load}(t) - \sum_{m=1}^{3} pil(m,t) + P_{sell}(t) + P_{ch}(t) + EV_{charge}(t) \tag{19}$$

The diesel generator operating constraints are upper and lower output limit constraints, climb rate constraints and start-stop constraints as shown in equations. (20)-(22) respectively.

$$u_{DG}(t) \cdot P_{DG,min} \leq P_{DG}(t) \leq u_{DG}(t) \cdot P_{DG,max} \tag{20}$$

$$-R_{DG} \leq P_{DG}(t) - P_{DG}(t-1) \leq R_{DG} \tag{21}$$

$$z_{DG}(t) \geq u_{DG}(t) - u_{DG}(t-1) \tag{22}$$

The electric vehicle constraints are shown in equations (23)-(25)

$$0 \leq EV_{charge}(t) \leq P_{EV,max} \tag{23}$$

$$\sum_{t=1}^{24} EV_{charge}(t) \cdot \eta_{charge} = Total_{EV\_Demand} \tag{24}$$

$$-\Delta_{EV} \leq EV_{charge}(t) - EV_{charge}(t-1) \leq \Delta_{EV} \tag{25}$$

Demand response load constraints are shown in equations (26)-(28)

$$0 \leq pil(m,t) \leq c_{il}(m) \cdot P_{load}(t) \tag{26}$$

$$\sum_{t=1}^{24} shiftload(t) = \sum_{t=1}^{24} P_{load}(t) \tag{27}$$

$$-0.2 \cdot P_{load}(t) \leq shiftload(t) - P_{load}(t) \leq 0.2 \cdot P_{load}(t) \tag{28}$$

The above model is a mixed integer linear programming problem (MILP), which is modeled using the YALMIP modeling tool in MATLAB and solved by the CPLEX optimization solver.

IV. EXAMPLE ANALYSIS

A microgrid in a northern park is simulated over a 24-hour dispatch cycle with a 1-hour time step. A time-of-use tariff is applied, as detailed in Table 1, and the load data is a typical bimodal load profile.

Table 1. Time-of-day tariff schedule.

Time period	Electricity price yuan/kwh
00:00-06:00; 22:00-24:00	0.3
06:00-07:00; 10:00-17:00	0.6
07:00-10:00; 17:00-22:00	0.95

As shown in Figure 2, in the baseline scenario, EVs concentrate on charging in the evening, leading to a significant increase in the evening peak load and the formation of a sub-peak at night. After the optimized dispatch, the load curve is obviously smoothed, achieving peak shaving and valley filling with fluctuation suppression. Demand response can cut 2-3 MW of peak load, but its effect on valley filling is limited. After the introduction of electric vehicles and energy storage, the peak load is reduced to 17 MW, and the low valley is raised to 12-14 MW, with significant regulation effect. The energy storage system charges in the low valley and discharges in the high peak, which not only optimizes the cost of purchased power but also improves the flexibility of system operation.

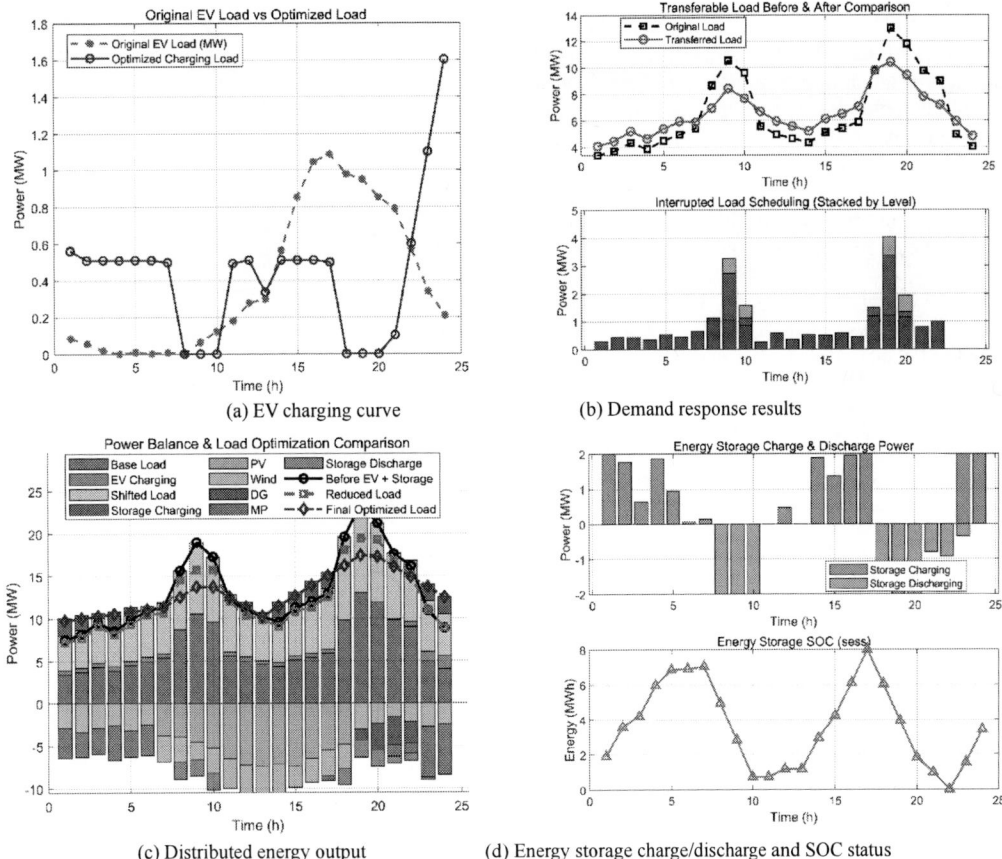

(a) EV charging curve

(b) Demand response results

(c) Distributed energy output

(d) Energy storage charge/discharge and SOC status

Figure 2. Simulation results of bimodal load.

To verify the validity of the experiment, two typical scenarios of single-peak load and double-peak load are set up for simulation and evaluation. As shown in Figure 3, the results show that under single-peak load, demand response reduces the peak load, and energy storage charging and discharging smooths out the fluctuations and makes the load curve smoother. In double-peak load, the peak is exacerbated by disorderly charging of electric vehicles, and staggered charging is achieved through optimized regulation to improve the load regulation capability. Compared with the existing research focusing on a single regulation means, this paper realizes the multi-scenario source-load-storage multi-distributed resource cooperative regulation, which can effectively deal with a variety of load characteristics, and provides a reference for distributed energy aggregation control.

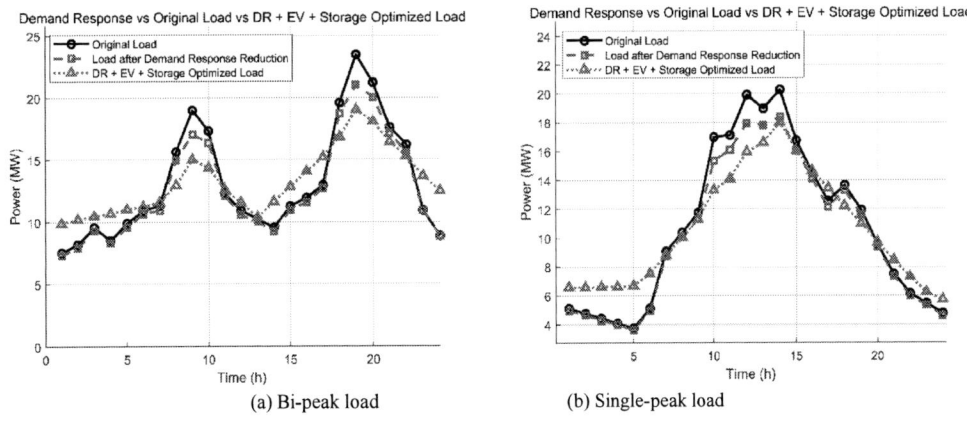

(a) Bi-peak load

(b) Single-peak load

Figure 3. Comparison of typical load optimization results.

979-8-3315-2490-6/25 $31.00 © 2025 IEEE

V. Conclusion.

This paper proposes a multi-distributed energy coordinated control method integrating demand response and EV behavior modeling. The approach reduces operating costs, cuts peak-valley differences by over 25.4%, and enhances system scheduling flexibility. The effectiveness of multi-energy co-optimization is verified through simulations. Future work will focus on incorporating renewable uncertainty and multi-agent participation for more accurate regulation.

References

[1] Wang Y. Z., Kang L. G., Zhang J., et al. Development history, typical forms and future trend of integrated energy system[J]. Acta Energiae Solaris Sinica, 2021, 42(08): 84-95.

[2] Kiehbadroudinezhad M, Merabet A, Abo-Khalil A G, et al. Intelligent and optimized microgrids for future supply power from renewable energy resources: A review[J]. Energies, 2022, 15(9): 3359.

[3] Cai L., Yang C.X., Li J.T., et al. Review of research and application on collaborative energy storage for electric vehicles accessed to active distribution network[J]. Journal of Chongqing Institute of Technology (Natural Science), 2024, 38(10): 212-220.

[4] Honarmand M E, Hosseinnezhad V, Hayes B, et al. An overview of demand response: From its origins to the smart energy community[J]. IEEE Access, 2021, 9: 96851-96876.

[5] Xie X. R., Ma N. J., Liu W., et al. Functions of energy storage in renewable energy dominated power systems: Review and prospect[J]. Proceedings of the CSEE, 2023, 43(01): 158-169.

[6] Zhang G, Wang W, Chen Z, et al. Modeling and optimal dispatch of a carbon-cycle integrated energy system for low-carbon and economic operation[J]. Energy, 2022, 240: 122795.

[7] Qiu B., Song S. X., Wang K., et al. Optimal operation of regional integrated energy system considering demand response and ladder-type carbon trading mechanism[J]. Proceedings of the CSU-EPSA, 2022, 34(05): 87-95+101.

[8] Patil S, Deshmukh S R. Development of control strategy to demonstrate load priority system for demand response program[C]//2019 IEEE International WIE Conference on Electrical and Computer Engineering (WIECON-ECE). IEEE, 2019: 1-6.

[9] Ogawa D, Kobayashi K, Yamashita Y. Blockchain-based optimization of energy management systems with demand response[C]//2020 IEEE 9th Global Conference on Consumer Electronics (GCCE). IEEE, 2020: 58-59.

Design of a WPT System in Implantable Medical Devices with Bluetooth-Assisted Alignment

Ziluo Ma*, Ran Ren‡†, Wenying Zhang§

*School of Foreign Languages, Beijing Forestry University, Beijing, China
maziluo1116@163.com

†Corresponding author

‡School of Informatics and Data Science, Hiroshima University, Hiroshima, Japan
c250014@hiroshima-u.ac.jp

§Faculty of Electric Power Engineering, Kunming University of Science and Technology, Kunming, China
kmzwying@sina.com

Abstract—**With the growing demand for implantable medical devices driven by an aging global population, power supply limitations have become a critical bottleneck. Conventional batteries restrict device lifespan and introduce risks associated with surgical replacement. Resonant magnetic coupling wireless power transfer (WPT) offers a promising solution, yet its efficiency is highly sensitive to coil alignment. This paper presents the development of a Bluetooth-assisted positioning system that integrates phase-based ranging (PBR) and round-trip time (RTT) technologies with Micro Processing Unit (MCU) processing to enhance alignment precision. Experimental results demonstrate improved charging performance in simulated biological conditions, thereby extending device longevity and expanding the applicability of WPT-enabled implants.**

Keywords-Wireless power transfer, Bluetooth ranging, Implantable medical devices, Coil alignment, Resonant magnetic coupling.

I. INTRODUCTION

As highlighted by recent market analyses [1], the demand for implantable medical devices has surged, fueled by demographic shifts and advancements in biomedical and materials sciences. Devices such as pacemakers, neurostimulators, and cochlear implants are increasingly utilized in clinical applications across cardiology, neurology, and endocrinology [2], [3]. Despite technological progress, these devices are often hindered by limited battery capacities, necessitating periodic surgical replacement, which imposes both financial and medical burdens.

Wireless power transfer (WPT), particularly via resonant magnetic coupling, has emerged as a viable method to address these limitations. However, effective implementation is contingent upon precise spatial alignment of the transmitter and receiver coils. Due to the dynamic and constrained nature of subcutaneous environments, achieving such alignment poses significant engineering challenges.

This study proposes a Bluetooth-based positioning mechanism that facilitates automatic alignment through accurate distance estimation, thus improving power transfer efficiency without manual intervention. The integration of phase-based ranging and RTT methods ensures robustness against environmental interference and supports real-time data processing via a MCU.

II. SYSTEM OVERVIEW

The proposed system comprises three main components: (1) a pair of resonant coils configured for efficient energy transfer; (2) a dual-chip Bluetooth module enabling precise spatial localization; and (3) a MCU, specifically the STM32F103C8T6, responsible for signal interpretation and alignment feedback control.

Unlike traditional manual alignment strategies, the system leverages wireless ranging technologies to dynamically adjust coil positioning. Bluetooth channel sensing, particularly PBR and RTT, is employed to enhance spatial resolution while maintaining low power consumption.

III. WIRELESS CHARGING SYSTEM DESIGN

A. Coil Design

Implantable devices impose strict constraints on component size and energy efficiency. Taking Medtronic's wireless cardiac pacemaker as a benchmark [4], device dimensions typically range from 43–53 mm in length and below 5 mm in thickness. Currently available cardiac pacemakers, such as those from Medtronic, typically feature dimensions of approximately 23 mm × 23 mm × 3 mm. As a result, the receiving coil must be confined within this compact form. Given the limited internal space and tightly integrated structure of pacemakers, this study adopts a planar coil configuration for the coupling mechanism. Although the external transmitting coil is less constrained in terms of size, an excessively large coil would result in only a small fraction of the generated magnetic field being effectively coupled to the internal receiver, with most of the electromagnetic energy being absorbed by surrounding biological tissue. This absorption poses potential safety risks due to electromagnetic radiation. Therefore, the transmitting coil is designed to be moderately enlarged to ensure adequate power transfer efficiency while minimizing adverse effects on human tissue, thus balancing system performance with biomedical safety considerations. The experimental platform finally uses spiral coils with the following specifications:

- Transmitting coil: 5 turns, outer diameter 44 mm, inner diameter 7 mm, spacing 3 mm, wire width 3 mm.

- Receiving coil: 5 turns, outer diameter 22 mm, inner diameter 4 mm, spacing 0.6 mm, wire width 1.5 mm.

979-8-3315-2490-6/25 $31.00 © 2025 IEEE

The system operates at a PWM frequency of approximately 200 kHz, optimized based on coil inductance and resistance characteristics, and must approve the standard of basic   restrictions for electromagnetic field exposure of the ICNIRP guidelines. Based on the simulated results of COMSOL Fig. 1, the radiation meets standards. All parameters are set to the input voltage 188 kHz and 5 V, determined by the actual spec of the coil.

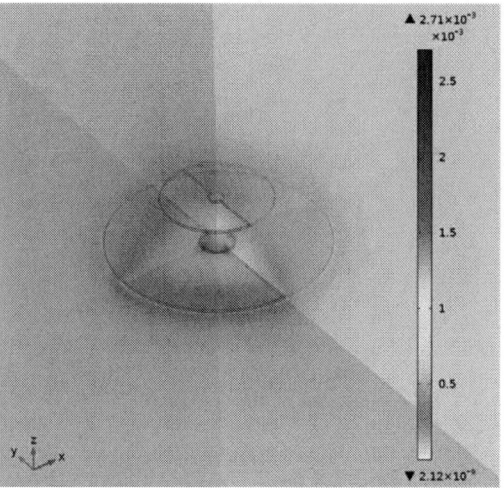

Fig. 1. COMSOL simulation

The equivalent circuit impedances are given as in formulas (1) and (2):

$$Z_S = \frac{R_s + j\omega L_s}{1 - \omega^2 L_s C_s + j\omega R_s C_s} \qquad (1)$$

$$Z_D = R_D + j\omega L_D + \frac{R_L}{1 + j\omega C_D R_L} \qquad (2)$$

B. Circuit Topology

Four typical resonant topologies exist for magnetic coupling WPT systems, "P" stands for parallel compensation, and "S" stands for series compensation: series-series (S-S), series- parallel (S-P), parallel-series (P-S), and parallel-parallel (P-P). As depicted in Fig. 2. The mathematical relations among inductance, capacity and angular velocity under different resonant topologies are as Table I.

TABLE I C_p AND C_S UNDER DIFFERENT RESONANT TOPOLOGIES

Resonant Topology	C_p	C_S
SS	$\dfrac{1}{\omega^2 L_p}$	$\dfrac{1}{\omega^2 L_s}$
PS	$\dfrac{L_p R_L^2}{\omega^2 L_p^2 R_L^2 + \omega^4 M^4}$	$\dfrac{1}{\omega^2 L_s}$
SP	$\dfrac{L_s}{\omega^2 L_p L_s - \omega^2 M^2}$	$\dfrac{1}{\omega^2 L_s}$
PP	$C_p = \dfrac{L_p - \dfrac{M^2}{L_s}}{\left(\dfrac{M^2 R_L}{L_s^2}\right)^2 + \omega^2\left(L_p - \dfrac{M^2}{L_s}\right)^2}$	$\dfrac{1}{\omega^2 L_s}$

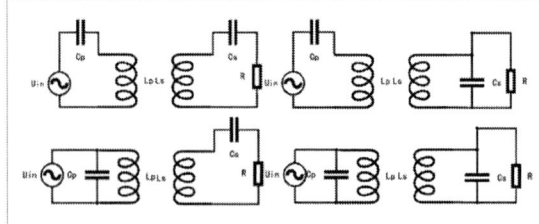

Fig. 2. Magnetic coupling topologies: S-S, S-P, P-S, P-P

According to the research result of W. Han [5], the P-P topology is selected due to its advantages in thermal performance and transfer efficiency in the implantable devices.

The explanations for each variable in the equation (1) are as the following:

- Z_s: Equivalent impedance of the source-side network

- R_s: Series resistance

- L_s: Series inductance

- C_s: Shunt capacitance

- ω: Angular frequency

- j: Imaginary unit

The explanations for each variable in the equation (2) are as the following:

- Z_D: Equivalent impedance on the load side

- R_D: Parasitic or series resistance on the load side

- L_D: Series inductance on the load side

- C_D: Shunt capacitance on the load side

- R_L: Load resistance

IV. POSITIONING SYSTEM REQUIREMENTS AND COMPARISONS

To be viable for biomedical applications, the positioning subsystem must satisfy the following criteria:

1) **Compact Form Factor and Low Power Demand:** Implantable devices necessitate minimal size and heat generation. The Bluetooth modules and control circuitry must function within strict power budgets to avoid thermal and electrical interference.

2) **Immunity to Electromagnetic Interference:** As power transmission inherently generates electromagnetic noise, the system utilizes distinct frequency bands for energy and signal transfer, thereby mitigating cross-talk and maintaining positional accuracy.

3) **External Real-Time Control:** A mechanical actuation module and external MCU are incorporated to process

localization data and execute dynamic coil alignment, ensuring stable operation under physiological conditions.

After years of research and development, several mainstream positioning methods have emerged in the industry, including magneto resistive sensor positioning, multi-soft coil matrix systems, and single-coil current pre-excitation technique [6]. Compared with these approaches, the Bluetooth-based positioning method proposed in this study offers a number of practical advantages, as outlined in Table II. Mag-

TABLE II COMPARISON OF POSITIONING METHODS FOR WIRELESS IMPLANT SYSTEMS

Method	Cost	Stability	Bio-safety
Magnetoresistive Sensor	50 USD	Moderate	Moderate
Multi-coil Matrix	90 USD	High	Low
Single Coil (Current Excitation)	10 USD	Low	Low
Proposed Bluetooth Method	1 USD	High	High

netoresistive sensor systems usually require additional coils to be placed inside the transmitting coil. This not only increases manufacturing complexity and cost but also leads to higher power consumption and potential instability. The multi-coil matrix method depends on flexible Polyethyleneterephthalate (abbreviated as PET) PCB coil arrays, which only perform optimally when custom-made. This customization significantly increases individual user costs and runs counter to the original goal of keeping implantable medical devices affordable. Additionally, all coils must operate simultaneously during detection, which creates stronger electromagnetic fields and raises safety concerns for the human body. This method is also more vulnerable to interference from nearby metal objects, which can affect positioning accuracy.

The single-coil current excitation approach requires the

system to send high-peak current pulses multiple times. To ensure stability, extra circuit redundancies are needed, which adds design complexity and increases the risk of malfunction. Like the previous method, it also causes considerable electromagnetic radiation, which is not ideal for patient safety. In this study, Bluetooth technology is operating on a frequency band that does not interfere with the electromagnetic coupling used for power transfer. Commercial Bluetooth modules are known for their ultra-low power consumption—as low as $37\mu A/MHz$—and are compact, often within a $1cm^2$ footprint, with costs generally under 1 USD. Thanks to continuous improvements in Bluetooth standards, the technology is expected to remain relevant and promising in the foreseeable future. More importantly, Bluetooth communication is minimally affected by biological tissues and does not pose risks such as protein denaturation caused by electromagnetic radiation.

IV. SYSTEM MECHANISM AND DESIGN

A. Bluetooth-Based Ranging Mechanism

The Phase-Based Ranging ranging process proceeds as follows:

1) The initiator transmits an unmodulated tone.

2) The responder receives the signal, measures phase, and sends a reply.

3) The initiator computes phase and time-based metrics to estimate distance.

This approach integrates Phase-Based Ranging (PBR) and Round-Trip Time (RTT) to provide an accurate and reliable positioning system. The combination of these techniques allows for a robust measurement process while reducing errors through validation steps. The MCU is responsible for controlling the data flow, filtering any anomalies, and ensuring the system remains stable for precise alignment. Bluetooth positioning is achieved through combined use of Phase-Based Ranging (PBR) and Round-Trip Time (RTT) techniques. In PBR, the phase offset between an incoming unmodulated tone and the local oscillator is used to compute the signal path length. RTT, in turn, estimates the distance by measuring the time delay for a signal's round trip.

1) Initialization and Pairing: The two Bluetooth chips automatically initiate and pair with each other to establish communication.

2) Phase-Based Ranging (PBR) - CS0 Phase: In this phase, the Bluetooth chips exchange information to determine the phase difference between their signals.

3) Phase-Based Ranging (PBR) - CS2 Phase: The system measures the path length based on the phase difference to estimate the distance.

4) Continuous Data Transmission: Once the distance is estimated, the system continues transmitting data to ensure stability and consistent readings.

5) Round-Trip Time (RTT) Phase: RTT is used to verify the distance calculated during the PBR phase by measuring the time taken for the signal to travel back and forth.

6) *Comparison Step:* The data is sent to the MCU for validation. If both data sets are consistent, the system calculates and outputs the average of the two as the final result. If not, the system will go back to the CS0 phase to start over.

The overall process is illustrated in Fig. 3.

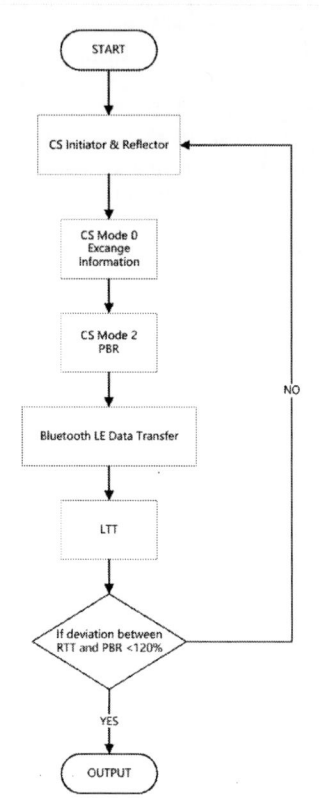

Fig. 3. System operation flowchart

B. System Hardware Configuration

An ideal wireless charging and positioning system should comprise several key components. In this study, we modified a commercially available solution to suit the design requirements, the whole system is as depicted in Fig 4.

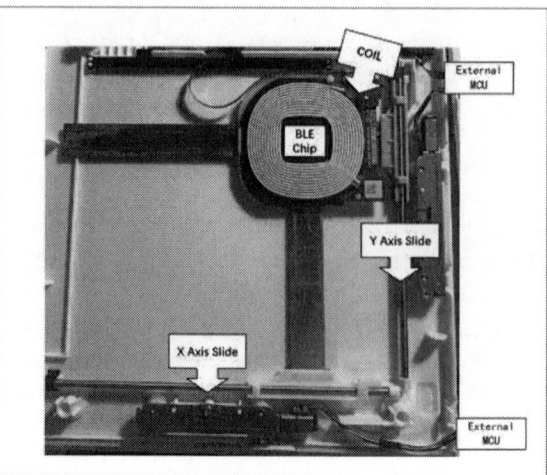

Fig. 4. Hardware System

At the core of the device is the transmitting coil, which serves as the primary energy transfer element. A magnetic sheet is affixed behind the coil to enhance magnetic field concentration and improve coupling efficiency. To avoid interference with magnetic flux, the Bluetooth module is deliberately positioned on the rear side of the coil. Placing it on the front would significantly degrade communication performance due to magnetic field obstruction.

The entire assembly is mounted on a dual-axis motorized translation stage. Both the X-axis and Y-axis motion modules,

as well as the Bluetooth module, are interfaced with an external MCU. The MCU collects positional data, processes alignment feedback, and executes control commands to adjust coil position dynamically. This configuration constitutes the complete transmitter-side architecture, enabling automated, high-precision spatial alignment with the implanted receiver.

V. CONCLUSION

This study presents a novel Bluetooth-enhanced positioning framework designed to optimize coil alignment in wireless power transfer (WPT) systems for implantable medical devices. By synergistically integrating phase-based ranging (PBR) and round-trip time (RTT) measurement methodologies with a MCU-based control architecture, the proposed system demonstrates exceptional alignment precision while maintaining robustness against electromagnetic interference. This innovation not only enhances power transfer efficiency but also substantially reduces the frequency of invasive battery replacement procedures, thereby advancing both operational reliability and patient safety. The implementation of a compact, energy-efficient Bluetooth module ensures minimal compromise to device miniaturization requirements and biological compatibility.

A key advantage lies in the system's adaptive capability to maintain performance integrity under dynamic clinical conditions, where misalignment between transmitting and receiving coils may occur due to physiological movements or environmental changes. This breakthrough represents a paradigm shift in WPT technology for medical implants, enabling the development of longer-lasting devices with improved energy management and reduced surgical intervention risks. Future work may focus on multisensory fusion and adaptive control algorithms to further elevate system performance.

REFERENCES

[1] cir.cn, "Research report on the current status and development trends of China's implantable medical device industry," *cir.cn*, 2024.

[2] C. X. Wang, "Research on wireless power supply for cardiac pacemakers based on magnetic negative metamaterials," Master dissertation *Liaoning Technical University*, vol. https://doi.org/10.27210/d.cnki.glnju.2023.000469, 2023.

[3] W. Yan and S. Yang, "Clinical applications and research progress of artificial hearts," *Journal of Precision Medicine*, vol. 37, no. 6, pp. 559–564, 2022.

[4] B. Nisarga, "Achieving high-precision, low-cost, and secure ranging using Bluetooth channel sensing," *Texas Instruments*, vol. ZHCADV 4, no. 6, pp. 1–6, 2024.

[5] W. Han, K. T. Chau, C. Jiang, and W. Liu, "Accurate position detection in wireless power transfer using magnetoresistive sensors for implant applications," *IEEE Transactions on Magnetics*, vol. 54, no. 11, p. 4001205, 2018.

[6] B. Zhang, Q. Chen, G. Ke, L. Xu, X. Ren, and Z. Zhang, "Coil positioning based on dc pre-excitation and magnetic sensing for wireless electric vehicle charging," *IEEE Transactions on Industrial Electronics*, vol. 68, no. 5, pp. 4125–4136, May 2021.

Research on Control Strategy of Single-Phase Inverter System Based on Osprey Optimization Algorithm

Huili Kang
Institute of Carbon Neutrality and New Energy, School of Electronics and Information
Hangzhou Dianzi University
Hangzhou China
2402216782@qq.com

Wensheng Yan*
Institute of Carbon Neutrality and New Energy, School of Electronics and Information
Hangzhou Dianzi University
Hangzhou China
wensheng.yan@hdu.edu.cn

Shengjie Ma
Institute of Carbon Neutrality and New Energy, School of Electronics and Information
Hangzhou Dianzi University
Hangzhou China
2423106124@qq.com

Lei Weng
Zhejiang Minglei Lithium Enerey Technology Co., ltd
Ningbo China
sales@minglei.com.cn

Zhen Zhang
Institute of Carbon Neutrality and New Energy, School of Electronics and Information
Hangzhou Dianzi University
Hangzhou China
2806352096@qq.com

Abstract—The control strategy plays an important role in the stable output of the inverter and the fast response to the load, and the resonant peak problem in the LC (Inductor Capacitor) filter often has a negative impact on the quality of the output waveform. In this paper, the single-phase inverter LC filter system is taken as the research object, and the operation and simulation analysis are carried out based on MATLAB platform. The results show that, compared with the traditional control strategy, the introduction of active damping modulation can effectively suppress the resonant peak. Combined with the outer loop PI (Proportional Integral) and inner loop QPR (Quasi Proportional Resonant) control strategy with optimized parameters of OOA (Osprey Optimization Algorithm), the output voltage and current waveforms of the system can respond quickly and remain stable when the load changes suddenly, and the waveform quality is significantly improved, the system THD (Total Harmonic Distortion) value is significantly reduced, indicating that the control strategy has excellent effect on reducing the harmonic distortion of the output waveform.

Keywords-Component; Double loop control; Inverter system; Harmonic suppression; OOA algorithm;

I. INTRODUCTION

With the rapid development of renewable energy and distributed generation technology, inverter, as the core equipment of power conversion, plays a vital role in modern power system. In order to meet the load demand for high-quality power, the output waveform of the inverter must have low harmonic content and high stability[1]. However, under the complex load mutation and nonlinear disturbance, the traditional control method is difficult to suppress the resonance peak of LC filter due to the lack of experience and dynamic response in parameter adjustment, which leads to the increase of THD of output waveform and the decrease of dynamic stability. How to design an intelligent control strategy with fast response and high precision has become a key challenge to improve the performance of the inverter.

Traditional PI control realizes voltage and current regulation through double closed-loop structure, but it needs to rely on passive damping when facing the resonance peak problem, which is easy to introduce additional loss. In recent years, intelligent algorithms, such as PSO (particle swarm optimization) and GA (genetic algorithm), have been used to optimize parameters, but there are still some shortcomings, such as PSO easily falling into local optimum[2] and GA converging slowly[3]. In the aspect of resonance suppression, active damping modulation reconstructs the control loop through virtual impedance, which provides a new idea for resonance peak suppression. However, the existing research on multi-focus single control link optimization lacks the cooperative design of algorithm selection and resonance suppression, which restricts the improvement of the overall performance of the system.

In this study, the LC filter system of single-phase inverter is taken as the object, and a double closed-loop cooperative control strategy based on the OOA is proposed. The innovations are as follows: 1) Combining outer-loop PI and inner-loop QPR control, the system transfer function is reconstructed by active damping modulation, and the resonance peak is effectively suppressed; 2) OOA algorithm is introduced to optimize PI parameters, which solves the premature convergence problem of traditional algorithms in high-dimensional parameter space, and

979-8-3315-2490-6/25 $31.00 © 2025 IEEE

the dynamic adaptive weight mechanism significantly improves the robustness under nonlinear disturbance. Experiments show that this strategy can reduce the THD of the system to 0.37%, reduce the steady-state response error by 57%, and still maintain high-precision tracking under sudden load change. This study provides a new technical path for intelligent control of power electronic equipment.

II. MATERIALS AND METHODS

A. Overview of Inverter Structure

Single-phase inverter usually consists of DC input source, full-bridge or half-bridge switching circuit, control circuit and output filter[4]. which converts DC into AC and eliminates high-frequency harmonics generated in the switching process through the filter.

B. Inverter Waveform Distortion

The single-phase inverter system designed in this paper adopts full-bridge inverter structure and the filter adopts LC structure. Before the analysis of the system and the design of the controller, it is necessary to establish a mathematical model for the single-phase inverter[5]. The full-bridge part is equivalent to a voltage source that periodically outputs positive and negative constant voltage with a period of 50 Hz, and obtain the state equivalent model as shown in Figure 1, with an inductance value of L=2mH, a capacitance value of C=5.6µF and a resonance frequency of f_r of 1504 Hz.

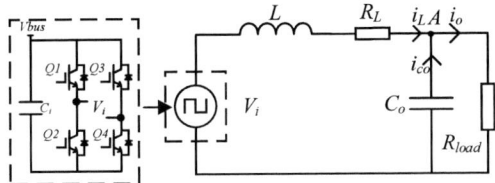

Fig. 1 Equivalent Model of Single-Phase Inverter

Using Kirchhoff's current theorem for point A in Figure 1 and Kirchhoff's voltage theorem for inductance L loop, we derive the state equation of LC filter circuit, and then get the following expression through frequency domain analysis:

$$\begin{cases} C_o V_o S = i_L - i_o \\ L i_L S = V_i - V_o - R_L i_L \end{cases} \quad (1)$$

After drawing the equivalent control block diagram of inverter system by Equation 1, we get the system transfer function as follows. Where Z_o is equivalent impedance.

$$G(S) = \frac{V_o}{V_i} = \frac{Z_o}{LC_o Z_o S^2 + (L + R_L C_o Z_o)S + Z_o + R_L} \quad (2)$$

$$Z_o = \frac{LS + R_L}{LC_o S^2 + C_o R_L S + 1} \quad (3)$$

According to Equation 1, the open-loop Bode diagram of LC filter at full load, half load and no load is drawn by MATLAB, and it is known that resonance peak will appear at no load.

C. Resonant Damping Control Strategy of Inverter

Because the resonance peak will appear in no-load and needs to be suppressed, the transfer function in no-load is

analyzed. At this time, Z_o tends to infinity, and the output current i_o tends to 0. The transfer function is equivalent to:

$$G(S) = \frac{V_o}{V_i} = \frac{1}{LC_o S^2 + R_L C_o S + 1} \quad (4)$$

Damping control strategies include passive control and active control[6]. Passive damping consumes the resonance energy of the filter by introducing appropriate damping elements (such as resistors) into the filter circuit, thus suppressing the resonance peak of the LC circuit. However, the added components in passive damping will produce power loss, which will affect the efficiency of the system. Therefore, active damping is adopted and the passive damping effect is equivalent through modulation control strategy[7].

Here, the passive damping mode of inductance series resistance is equivalent to the active damping control mode, and the control block diagram is shown in Figure 2, where k_r is the resistance equivalent parameter of the passive damping control strategy of inductance series resistance.

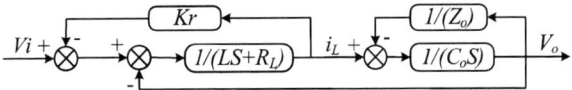

Fig. 2 LC Filter Active Damping Control Block Diagram

The Bode diagram of the comparison between active damping control strategy and undamped control strategy at no-load is shown in Figure 3. By comparing them, we can see that the resonance peak at the resonance frequency is obviously suppressed.

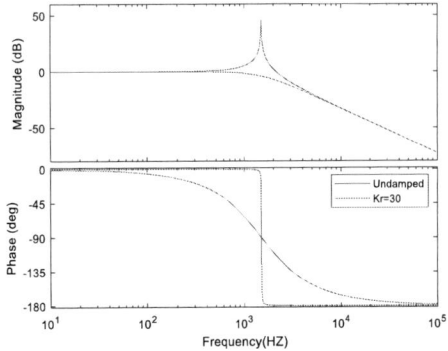

Fig. 3 Bode Diagram of Active Damping and Undamped under No Load

D. Double Closed Loop Control

The inverter system adopts the double closed-loop control strategy of voltage outer loop PI control and current inner loop QPR control. The voltage outer-loop PI control is mainly responsible for the stability of the system and the load adjustment, so as to ensure the stability of the inverter output voltage. The current inner loop QPR control focuses on the control of instantaneous current to ensure that the inverter output current accurately follows the set value.

The control block diagram of voltage outer loop and current inner loop with double closed loops is shown in Figure 4.

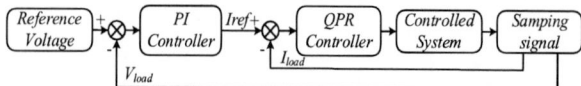

Fig. 4 Voltage and Current Double Closed-Loop Control Block Diagram

PI control transfer function is:

$$G_{PI}(S) = K_p + \frac{K_i}{S} \text{ [8]} \qquad (5)$$

OPR control transfer function is:

$$G_{QPR}(S) = K_p + K_r \frac{2K_r W_c S}{S^2 + 2W_c S + W_o^2} \text{ [9]} \qquad (6)$$

E. Intelligent Algorithm Optimization Based on OOA

The parameters in the PI-QPR double closed loop used in this system are often difficult to find the optimal solution by empirical manual debugging, so an intelligent algorithm is introduced to optimize the parameter selection in the inverter control strategy. The OOA is a swarm intelligence optimization algorithm based on osprey fishing behavior. The core idea is to search the optimal solution by simulating the fishing behavior of the Osprey. The introduction of the OOA can automatically adjust the key parameters of the controller according to the real-time running state of the system, so as to optimize the system performance.

The formula of osprey exploration stage is:

$$X_{i,j}1 = X_{i,j} + R_{i,j}(SF_{i,j} - I_{i,j}X_{i,j}) \qquad (7)$$

The formula of Osprey development stage is:

$$X_{i,j}2 = X_{i,j} + \frac{lb_j + R_{i,j}(ub_j - lb_j)}{t} \qquad (8)$$

Where $R_{i,j}$ is random numbers between [0,1], $SF_{i,j}$ is randomly selected fish positions, $I_{i,j}$ is random numbers between sets [1], lb_j is the lower limit of j dimension, ub_j is the upper limit of j dimension, i=1,2,3...,N, j = 1,2,3 ...,M, t=1,2,3...,T, N is the total population of osprey, M is the dimension, and T is the number of iterations.

The osprey randomly finds the position of one of the fish and attacks it, and uses formula(7) to calculate the new position of the corresponding osprey. If the new position is better, replace the previous position of the osprey. Then take it to a suitable place to eat, and use formula (8) to calculate the new position of the corresponding osprey. If the new position is better, replace the previous position of the osprey.

III. RESULTS AND DISCUSSION

Use MATLAB/Simulink to build the simulation model of inverter system, as shown in Figure 5. The model control adopts voltage outer loop PI control and current inner loop QPR control, and the OOA optimizes the PI parameters. In the simulation of OOA, the population number is 30, the iteration number is 30, the upper and lower limits of K_p are [0.2, 0.8], and the upper and lower limits of K_i are [2000, 8000]. The optimal solution is obtained by taking the integral of absolute value of time times error as the objective function, and the final parameters K_p is 0.5497 and K_i is 5921.4.

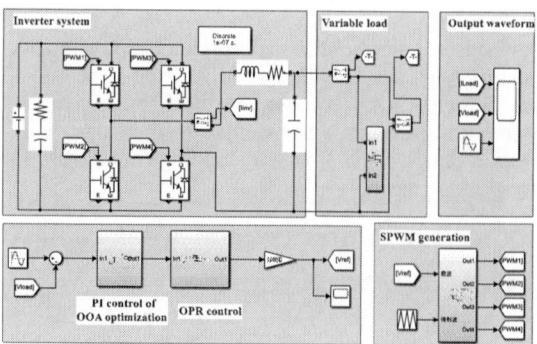

Fig. 5 Simulation Model of Inverter System

The waveform obtained by system simulation under steady state is shown in Figure 6. We can see that the output voltage of PI-PI double closed loop in Figure 6(a) and the standard sine wave are not coincident at the peak. Figure 6(b) shows that there is a large output error of -20.8V at the beginning, and the output error is 6.65V after stabilization. In Figure 6(c), the output voltage of the PI-QPR double closed loop and the standard sine wave still coincide at the peak after OOA optimizes the PI parameters, and the overall coincidence degree is high. As shown in Figure 6(d), there is no big error in the whole process, and the output error is 2.85V after stabilization.

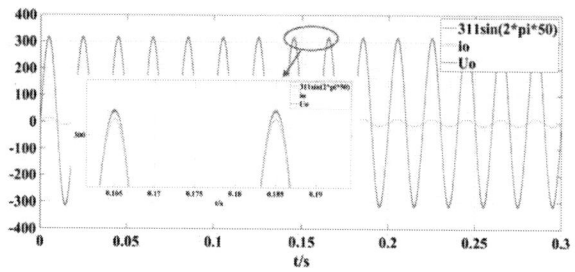

Fig. 6 (a) Combination Diagram of PI-PI Double Closed-Loop Output Voltage, Output Current and Standard Sine Wave

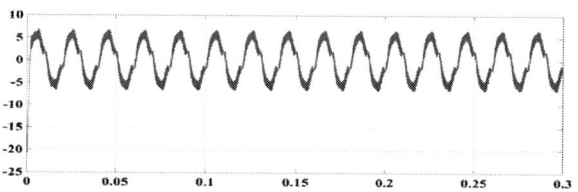

Fig. 6 (b) PI-PI Double Closed-loop Output Error Image

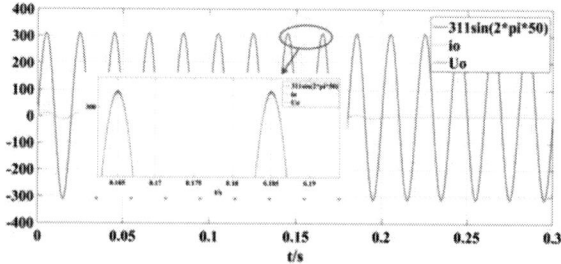

Fig. 6(c) Combination Diagram of PI-QPR Double Closed-Loop Output Voltage, Output Current and Standard Sine Wave After OOA

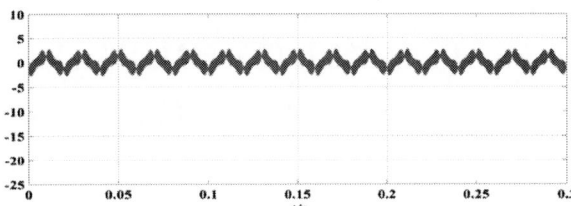

Fig. 6 (d) PI-QPR Double Closed-Loop Output Error Image After OOA

Fig. 6 Output Waveform Diagram and Output Error Comparison Diagram

The waveform obtained by system simulation under dynamic conditions is shown in Figure 7. The system initially works in full-load mode, switches to half-load mode at 0.1s and switches to no-load mode at 0.2s. We can see that the overall output error of the PI-PI double-closed loop in Figure 7(a) decreases with the decrease of load, and there is an oscillation peak at the sudden change of load of 0.1s and 0.2s, which makes the output unstable at this moment, and the output errors are 6.65V, 5.70V and 4.80V respectively. In Figure 7(b), the overall output error of PI-QPR double closed-loop decreases with the decrease of load, and the output errors are 2.85V、2.15V、1.80V respectively.

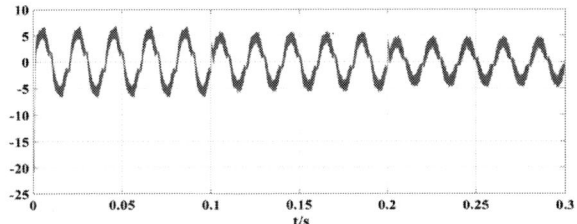

Fig. 7 (a) PI-PI Double Closed-Loop Output Error Image

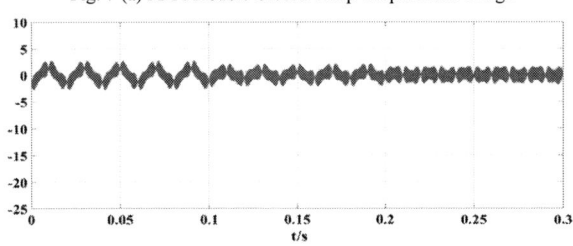

Fig. 7 (b) PI-QPR Double Closed-Loop Output Error Image After OOA

Fig. 7 Comparison Chart of Output Error in Dynamic State

To sum up, the effects of steady-state and dynamic performance show that the inverter has high stability and fast response ability after the control strategy of this system.

Figure 8 shows the total harmonic distortion (THD) diagram of the output current waveform. We can see that the THD is 0.43% in Figure 8(a) and 0.37% in Figure 8(b). It shows that the THD value is significantly reduced under the condition of active damping modulation and PI-QPR double closed-loop control modulation with OOA optimized PI parameters, which shows that the control strategy has a good effect in reducing harmonic distortion of output waveform.

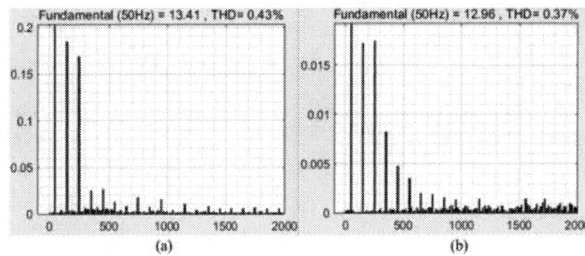

Fig. 8 Comparison Diagram of Output Current THD   (a)PI-PI Double Closed-Loop Output Current THD   (b) PI-QPR Double Closed-Loop Output Current THD After OOA

IV. CONCLUSIONS

The control of active damping not only weakens the influence of resonance peak of LC filter when there is no load, but also avoids the extra power loss of the system under the passive damping control strategy. The optimization of voltage outer-loop PI control and current inner-loop QPR control significantly improves the tracking performance of inverter system. After further introducing the Osprey algorithm to optimize the PI control parameters, the system can maintain the best response speed and accuracy under various working conditions, thus improving the stability and robustness of the inverter. On the whole, the inverter system adopting the optimized control strategy in this paper has stronger load adaptability, higher efficiency and longer stable operation time, and can meet higher requirements of application scenarios.

REFERENCES

[1] Wei Y, Jiang Z, Lv T, et al. A Novel Single-Stage Boost Single-Phase Inverter and Its Composite Control Strategy to Suppress Low-Frequency Input Ripples[J]. Energies, 2024, 17(17).

[2] Abdolrasol MGM, Ayob A, Mutlag A H, et al. Optimal fuzzy logic controller based PSO for photovoltaic system [J]. Energy Reports, 2023, 9: 427-34.

[3] Ali M, Tariq M, Upadhyay D, et al. Genetic Algorithm Based PI Control with 12-Band Hysteresis Current Control of an Asymmetrical 13-Level Inverter [J]. Energies, 2021, 14(20).

[4] Pan Z, Chen X, Geng G, et al. Second harmonic current reduction of dual active bridge converter under dual-phase-shift control in two-stage single-phase inverter for residential energy storage system[J]. Journal of Energy Storage, 2024, 103.

[5] She Y, Huo X, Tong X, et al. Multi-Sampling Rate Finite Control Set Model Predictive Control and Adaptive Method of Single-Phase Inverter[J]. Electronics, 2023, 12(13).

[6] Ying G, Zeng J, Wu M, et al. An enhancement method ensuring robust passive impedance of LCL-type grid connected converter[J]. Electric Power Systems Research, 2025, 243.

[7] Hou T, Jiang Y, Cai Z. Study on the Resonance Characteristics and Active Damping Suppression Strategies of Multi-Inverter Grid-Connected Systems Under Weak Grid Conditions[J]. Energies, 2024, 17(23).

[8] Kashfi R, Balochian S, Alishahi M. Design of a optimal robust adaptive neural network-based fractional-order PID controller for H-bridge single-phase inverter[J]. Applied Soft Computing, 2024, 166.

[9] Sang Y K, Jun B K, Wook D K, et al. Design method of PR Controller Considering Active Damping based on Output Impedance of Single Phase Full Bridge Inverter[J]. Journal of the Korean Institute of Illuminating and Electrical Installation Engineers, 2020, 34(1): 18-2.

2025 International Conference on Advanced Energy Systems and Power Electronics (AESPE 2025)

Open-Circuit-Fault Diagnosis and Fault-Tolerant Strategy for Flying Capacitor Dual-Active-Bridge DC-DC Converter

Shunjie Jiang [1,a]
[1] School of Electrical Engineering, Guangxi University, Nanning, Guangxi, China
[a]e-mail: jiangshunjie@163.com

Wusheng Shi [2,b*]
[2] School of Electrical Engineering, Guangxi University, Nanning, Guangxi, China
[*b]Corresponding author's e-mail: shiwusheng@gxu.edu.cn

Dalong Hu [3,c]
[3] School of Electrical Engineering, Guangxi University, Nanning, Guangxi, China
[c]e-mail: 1239544946@qq.com

Yuxin Zhou [4,d]
[4] School of Electrical Engineering, Guangxi University, Nanning, Guangxi, China
[d]e-mail: yxzhou_04@163.com

Wen Guo [5,e]
[5] School of Electrical Engineering, Guangxi University, Nanning, Guangxi, China
[e]e-mail: guowen_neu@163.com

Abstract-The flying capacitor dual-active-bridge (FC-DAB) dc-dc converter is a promising dc-dc converter for medium-voltage and high-voltage applications. However, the open-circuit switching fault (OCSF) of the FC-DAB converter may result in various negative effects such as transformer voltage distortion, inductor current dc bias. Therefore, fault diagnosis and fault-tolerant control strategy are essential. Fault diagnosis merely based on the polarity of inductor current or the midpoint voltages of the bridge arms are not applicable for the FC-DAB converter. Moreover, FC-DAB converter has more working states which provides potential for fault-tolerant operation. To address these issues, this paper proposes a fault diagnosis and fault-tolerant control strategy for the FC-DAB converter. In the proposed fault diagnose strategy, the average values of the midpoint voltages of the bridge arms and the low-potential side voltages of flying capacitors are utilized to identify the faulty switch. Based on blocking the complementary switch, symmetry operational characteristics can be restored with the proposed fault-tolerant strategy. The proposed fault diagnosis and fault-tolerant control strategy are verified by simulation.

Keywords-Flying Capacitor Dual-Active-Bridge, Open-Circuit Switching Fault, Fault Diagnosis, Fault-Tolerant Control, Reliability

I. INTRODUCTION

Due to advantages of galvanic isolation, bidirectional power transmission and high power density, dual-active-bridge (DAB) dc-dc converter has a wide application prospects in renewable energy systems, dc microgrids, and energy storage systems [1]. The three-level flying capacitor topology has the characteristic of low voltage stress, the DAB converter constructed based on the three-level flying capacitor topology is suitable for medium-voltage and high-voltage applications [2]. Furthermore, the flying capacitor dual-active-bridge (FC-DAB) enables increased control degrees of freedom through different combination of switching states, thereby improving flexibility.

Ensuring the reliability of power electronics converters is critical. According to previous studies, switching devices failure and driving failure are among the common failure types in the power electronics devices [3]. As a result, open-circuit switching fault (OCSF) of the converter has attracted considerable attention in the field of reliability research. When an OCSF occurs on the FC-DAB converter, operational characteristics will be asymmetrical, resulting in transformer voltage distortion, dc bias, and other issues, adversely affect the stable performance of the converter. Thus, fault diagnosis and fault-tolerant control are the means to ensure the reliable operation of FC-DAB converter after an OCSF occurs.

Various fault diagnosis methods have been proposed for both two-level DAB converters and three-level DAB converters. Davoodi et al. proposed a fault diagnosis methods based on the polarity of the dc component of the inductor current [4]. In the research conducted by Airabella et al., transformer voltages were utilized for fault diagnosis. In order to distinguish the faulty switch between the symmetric position, in the research conducted by Zheng et al., the average midpoint voltages of the bridge arms were applied [6]. Song et al. proposed a fault diagnosis method based on the mean values and the duty cycles of the midpoint voltages of the bridge arms for the three-level neutral-point-clamped dual-active-bridge (NPC-DAB) converter [7]. However, these methods are not applicable for the FC-DAB converter. Under certain conditions, when an OCSF occurs within one of these two switches on the low-potential side or high-potential side of a bridge arm, the polarity of inductor current or the distortion on midpoint voltages of the bridge arms are the same. Thus, the fault diagnosis for the FC-DAB converter needs to be addressed.

After the faulty switch is identify, an appropriate fault-tolerant control strategy can effectively mitigate the negative effects caused by OCSF. Zhao et al. proposed a fault-tolerant

979-8-3315-2490-6/25 $31.00 © 2025 IEEE   58

method for the secondary side of the two-level DAB converter based on blocking the gate-driving signals for the entire arm of the faulty switch locates [8]. Wu et al. proposed a fault-tolerant method based on blocking the complementary switch for the primary side inner switch or the secondary side switch of the NPC-DAB converter [9]. However, for the FC-DAB converter, the symmetry operational characteristics can be restored based on certain combination of switching states during postfault operation, regardless of the position of the faulty switch.

In order to address the issues, this paper proposes a fault diagnosis and fault-tolerant control strategy for the FC-DAB converter. This paper is organized as follows. Section II presents the topology and operation principle of the FC-DAB converter. Section III analyses the dynamic characteristics when the OCSF occurs on the FC-DAB converter, and proposes the fault diagnosis strategy. Section IV proposes the fault-tolerant strategy, and illustrates the operating principle. Section V verifies the effectiveness of the proposed fault diagnosis and fault-tolerant control strategy by the simulation results. Finally, Section VI concludes this paper. The logical flow chart of the entire study is shown in Scheme 1.

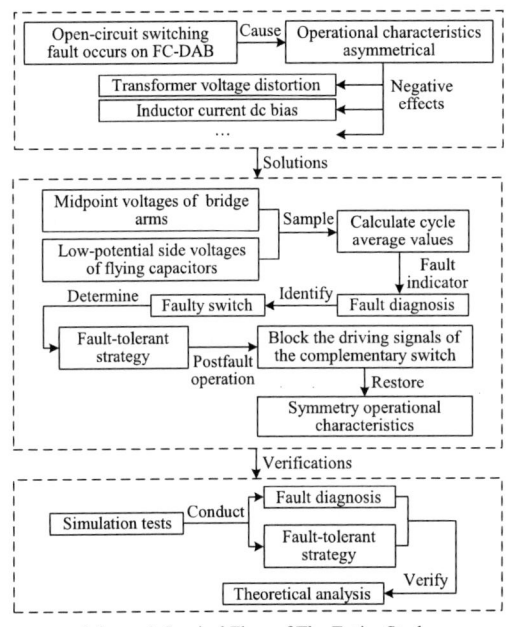

Scheme 1. Logical Flow of The Entire Study.

II. BASIC STRUCTURE AND NORMAL OPERATION MODE OF FC-DAB

A. Basic Structure of FC-DAB

The topological structure of FC-DAB is shown in Fig. 1. V_1 and V_2 are the input and output voltages. v_{ab} and v_{cd} are the ac output voltages of the primary side bridge and the secondary side bridge. L is the auxiliary inductance. i_L is the inductor current. T is the high-frequency transformer, and the turn ratio is defined as $n{:}1$. C_{f11} and C_{f12} are the flying capacitors of the primary side bridge. $S_{11}{\sim}S_{18}$ and $D_{11}{\sim}D_{18}$ are the switches and the antiparallel diodes of the primary side bridge, respectively. C_{f21} and C_{f22} are the flying capacitors of the secondary side bridge. $S_{21}{\sim}S_{28}$ and $D_{21}{\sim}D_{28}$ are the switches and the antiparallel diodes of the secondary side bridge, respectively.

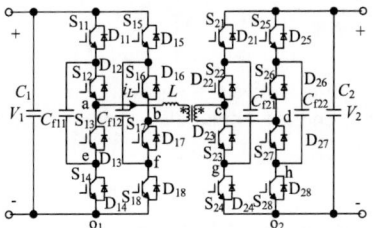

Fig. 1. Topological Structure of FC-DAB.

B. Normal Operation Mode of FC-DAB

The waveforms of the gate-driving signals, the voltages, and the inductor current with the single-phase-shift (SPS) control scheme under the condition buck mode with forward power flow is shown in Fig. 2. In this paper, f_s is the switching cycle, and a half of the switching cycle $T_{hs}=1/(2f_s)$. d is the phase-shift ratio between the primary side and the secondary side bridges, and d is in the range of $(-1, 1)$. k is the voltage conversion ratio, $k=V_1/nV_2$.

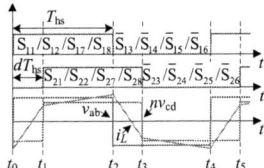

Fig. 2. Waveforms of FC-DAB With Single-Phase-Shift Control.

In Fig. 2, t_0 is defined as zero time, then $t_1{\sim}t_4$ can be expressed as $t_1=dT_{hs}$, $t_2=T_{hs}$, $t_3=(1+d)T_{hs}$, $t_4=2T_{hs}$. In the steady state, i_L exhibits the property of waveform periodic symmetry, thus the current values in $t_0{\sim}t_2$ can be calculated as follows.

$$
\begin{cases}
i_L(t_0) = \dfrac{nV_2}{4Lf_s}(1-k-2d) \\[2mm]
i_L(t_1) = \dfrac{nV_2}{4Lf_s}(1-k+2kd) \\[2mm]
i_L(t_2) = \dfrac{nV_2}{4Lf_s}(k-1+2d)
\end{cases}
\tag{1}
$$

The transmission power P related to d can be expressed as:

$$
P = \frac{1}{T_{hs}}\int_0^{T_{hs}} V_1(t)i_L(t)\,dt = \frac{nV_1V_2T_{hs}}{L}d(1-d)
\tag{2}
$$

From (2), when d is 0.5, P reaches its maximum. The maximum transmission power P_{max} can be expressed as:

$$
P_{max} = \frac{nV_1V_2T_{hs}}{4L}
\tag{3}
$$

The unified transmission power P^* can be expressed as:

$$
P^* = \frac{P}{P_N} = 4d(1-d)
\tag{4}
$$

where $P_N=nV_1V_2T_{hs}/4L$ is the maximum transmission power of the FC-DAB converter under SPS control in normal operation state.

979-8-3315-2490-6/25 $31.00 © 2025 IEEE

III. OPEN-CIRCUIT FAULT MODE ANALYSIS AND FAULT DIAGNOSIS OF FC-DAB

To simplify the analysis, assume that only one switch is in the fault state, and the antiparallel diode of the faulty switch operates normally.

A. Primary Side Bridge Open-Circuit Switching Fault Mode

The postfault operating modes is related to k and d. Only the condition $1<k<2$ and $(k-1)/2k<d<(-k^2+3k+2)/(6k+4)$ is presented, with remaining conditions following the similar approaches. Assuming that the fault occurs at the time corresponding to the rising edge of the gate-driving signal for the switch in normal operation state.

The transient waveforms when an OCSF occurs on S_{11} in shown in Fig. 3(a). In this paper, v_a is the voltage between a and o_1. v_b is the voltage between b and o_1. v_{ef} is the voltage between e and f. Assuming that the fault occurs at t_4. In $[t_4, t_5)$, the current flows through D_{18}, D_{17}, D_{12} and D_{11}, as shown in Fig. 4(a). This state is the same as normal operation. In this state, $v_{ab}=V_1$, $v_a=V_1$, $v_b=0$, $v_{ef}=V_1/2$. Because $V_1>-nV_2$, i_L keep increasing. Then, i_L changes from negative to positive at t_5. In $[t_5, t_6)$, due to the positive i_L cannot flow through the faulty switch S_{11}, the current flow through D_{14}, C_{f11}, S_{12}, S_{17} and S_{18}, as shown in Fig. 4(b). This state does not present in normal operation. In this state, v_{ab} and v_a reduce from V_1 to $V_1/2$, $v_b=0$, v_{ef} reduce from $V_1/2$ to 0. Because $V_1/2>-nV_2$, i_L keep increasing. In $[t_6, t_7)$, the current conduction path of the primary side bridge is the same as $[t_5, t_6)$. In this state, because $V_1/2<nV_2$, i_L keep decreasing. Then, i_L reduce to zero at t_7. In $[t_7, t_8)$, because S_{11} breaks down, $V_1/2<nV_2$ and $V_1>nV_2$, current flow is interrupted in this state. Thus, v_{ab} and v_a become nV_2, $v_b=0$, and v_{ef} become $nV_2-V_1/2$ in this state.

The transient waveforms when an OCSF occurs on S_{12} in shown in Fig. 3(b). Assuming that the fault occurs at t_4. In $[t_4, t_5)$, the current flows through D_{18}, D_{17}, D_{12} and D_{11}, the current path is the same as normal operation. In this state, $v_{ab}=V_1$, $v_a=V_1$, $v_b=0$, $v_{ef}=V_1/2$. Because $V_1>-nV_2$, i_L keep increasing. Then, i_L changes from negative to positive at t_5. In $[t_5, t_6)$, due to the positive i_L cannot flow through the faulty switch S_{12}, the current flow through S_{11}, C_{f11}, D_{13}, S_{17} and S_{18}, as shown in Fig. 4(c). This state does not present in normal operation. In this state, v_{ab} and v_a reduce from V_1 to $V_1/2$, $v_b=0$, v_{ef} reduce from $V_1/2$ to 0. Because $V_1/2>-nV_2$, i_L keep increasing. In $[t_6, t_7)$, the current conduction path of the primary side bridge is the same as $[t_5, t_6)$. In this state, because $V_1/2<nV_2$, i_L keep decreasing. Then, i_L reduce to zero at t_7. In $[t_7, t_8)$, because S_{12} breaks down, $V_1/2<nV_2$ and $V_1>nV_2$, current flow is interrupted in this state. Thus, v_{ab} and v_a become nV_2, $v_b=0$, and $v_{ef}=V_1/2$ in this state.

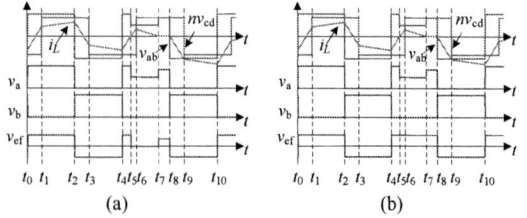

Fig. 3. Transient Waveforms When an OCSF Occurs on The Primary Side Bridge. (a) OCSF Occurs on S_{11}, (b) OCSF Occurs on S_{12}.

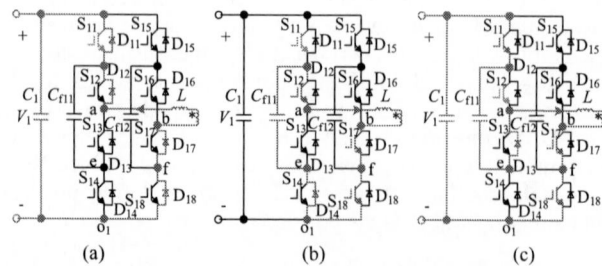

Fig. 4. Current Conduction Path of The Primary Side Bridge. (a) $[t_4, t_5)$ of S_{11} Fault, (b) $[t_5, t_6)$ of S_{11} Fault, (c) $[t_5, t_6)$ of S_{12} Fault.

The transient waveforms when an OCSF occurs on S_{13}~S_{18} in shown in Fig. 5. The working states of S_{13}~S_{18} fault are complementary to S_{11} or S_{12} fault.

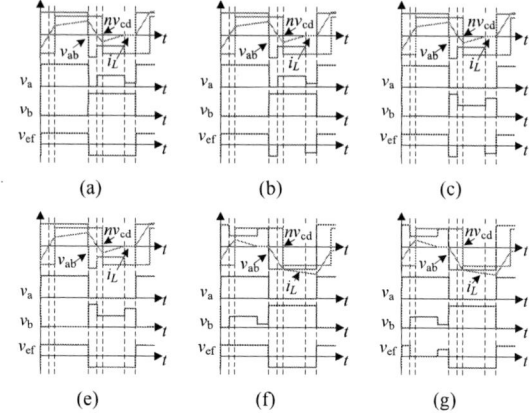

Fig. 5. Transient Waveforms When an OCSF Occurs on The Primary Side Bridge. (a) OCSF Occurs on S_{13}, (b) OCSF Occurs on S_{14}, (c) OCSF Occurs on S_{15}, (d) OCSF Occurs on S_{16}, (e) OCSF Occurs on S_{17}, (f) OCSF Occurs on S_{18}.

B. Secondary Side Bridge Open-Circuit Switching Fault Mode

Similarly, only the condition $k>1$ and $(k-1)/2k<d<0.5$ is presented, with remaining conditions following the similar approaches. Assuming that the fault occurs at the time corresponding to the rising edge of the gate-driving signal for the switch in normal operation state.

The transient waveforms when an OCSF occurs on S_{21} in shown in Fig. 6(a). In this paper, v_c is the voltage between c and o_2. v_d is the voltage between d and o_2. v_{gh} is the voltage between g and h. Assuming that the fault occurs at t_4. In $[t_4, t_5)$, the current flows through D_{28}, D_{27}, D_{22} and D_{21}, as shown in Fig. 7(a). This state is the same as normal operation. In this state, $v_{cd}=V_2$, $v_c=V_2$, $v_d=0$, $v_{gh}=V_2/2$. Because $V_1>nV_2$, i_L keep increasing. In $[t_5, t_6)$, because $-V_1<nV_2$, i_L keep decreasing. In this state, the current conduction path of the secondary side bridge is the same as $[t_4, t_5)$. Then, i_L changes from positive to negative at t_6. In $[t_6, t_7)$, due to the negative i_L cannot flow through the faulty switch S_{21}, the current flow through D_{24}, C_{f21}, S_{22}, S_{27} and S_{28}, as shown in Fig. 7(b). This state does not present in normal operation. In this state, v_{cd} and v_c reduce from V_2 to $V_2/2$, $v_d=0$, v_{gh} reduce from $V_2/2$ to 0. Because $-V_1<nV_2/2$, i_L keep decreasing.

The transient waveforms when an OCSF occurs on S_{22} in

shown in Fig. 6(b). Assuming that the fault occurs at t_4. In $[t_4, t_5)$, the current flows through D_{28}, D_{27}, D_{22} and D_{21}, as shown in Fig. 7(a). This state is the same as normal operation. In this state, $v_{cd}=V_2$, $v_c=V_2$, $v_d=0$, $v_{gh}=V_2/2$. Because $V_1>nV_2$, i_L keep increasing. In $[t_5, t_6)$, because $-V_1<nV_2$, i_L keep decreasing. In this state, the current conduction path of the secondary side bridge is the same as $[t_4, t_5)$. Then, i_L changes from positive to negative at t_6. In $[t_6, t_7)$, due to the negative i_L cannot flow through the faulty switch S_{22}, the current flow through S_{21}, C_{f21}, D_{23}, S_{27} and S_{28}, as shown in Fig. 7(c). This state does not present in normal operation. In this state, v_{cd} and v_c reduce from V_2 to $V_2/2$, $v_d=0$, and $v_{gh}=V_2/2$. Because $-V_1<nV_2/2$, i_L keep decreasing.

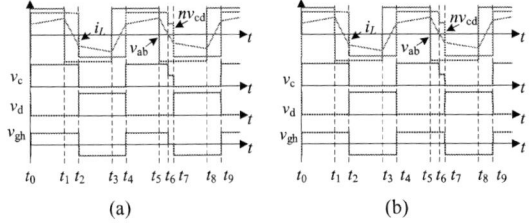

(a)   (b)

Fig. 6. Transient Waveforms When an OCSF Occurs on The Secondary Side Bridge. (a) OCSF Occurs on S_{21}, (b) OCSF Occurs on S_{22}.

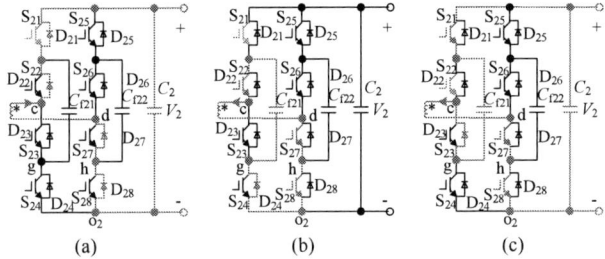

(a)   (b)   (c)

Fig. 7. Current Conduction Path of The Secondary Side Bridge. (a) $[t_4, t_5)$ of S_{21} Fault, (b) $[t_6, t_7)$ of S_{21} Fault, (c) $[t_6, t_7)$ of S_{22} Fault.

The transient waveforms when an OCSF occurs on S_{23}~S_{28} in shown in Fig. 8. The working states of S_{23}~S_{28} fault are complementary to S_{21} or S_{22} fault.

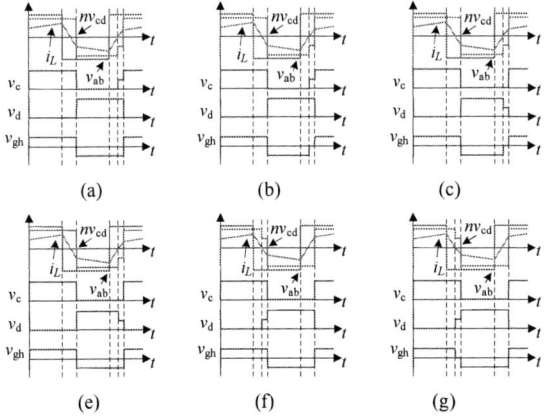

(a)   (b)   (c)

(e)   (f)   (g)

Fig. 8. Transient Waveforms When an OCSF Occurs on The Secondary Side Bridge. (a) OCSF Occurs on S_{23}, (b) OCSF Occurs on S_{24}, (c) OCSF Occurs on S_{25}, (d) OCSF Occurs on S_{26}, (e) OCSF Occurs on S_{27}, (f) OCSF Occurs on S_{28}.

C. Fault Diagnosis Strategy

According to the theoretical analysis of the fault modes, due to the current cannot flow through the faulty switch in certain state, the current conduction path will be changed, and it will cause the waveforms of current and voltage distortion. It is noticed that the distortion of i_L, v_{ab}, v_{cd}, v_a, v_b, v_c, and v_d is the same when an OCSF occurs on one of these two switches on the low-potential side or high-potential side of a bridge arm in certain condition (e.g., S_{11} and S_{12} in Fig. 3), so the faulty switch cannot be identified merely base on those variables. Therefore, v_a, v_b, v_c, v_d, v_{ef} and v_{gh} are measured to identify the faulty switch.

In this paper, v_{a_avg}, v_{b_avg}, v_{c_avg}, v_{d_avg}, v_{ef_avg} and v_{gh_avg} are the duty cycles of the average value of v_a, v_b, v_c, v_d, v_{ef} and v_{gh}, respectively. v_{a_avg}, v_{b_avg}, and v_{ef_avg} are utilized to identify the faulty switch for the primary side bridge, and v_{c_avg}, v_{d_avg}, and v_{gh_avg} are utilized to identify the faulty switch for the secondary side bridge. Under normal operation condition, the working states of FC-DAB converter exhibit the property of periodic symmetry, so v_{a_avg}, v_{b_avg}, v_{c_avg}, v_{d_avg}, v_{ef_avg} and v_{gh_avg} can be calculated as $v_{a\_avg}=V_1/2$, $v_{b\_avg}=V_1/2$, $v_{ef\_avg}=0$, $v_{c\_avg}=V_2/2$, $v_{d\_avg}=V_2/2$, and $v_{gh\_avg}=0$. When an OCSF occurs on S_{11}, during postfault operation, when the current conduction path changed instead of flowing through the S_{11} under normal condition, the amplitude of v_a and v_{ef} will reduce, causes v_{a_avg} to be less than $V_1/2$ and v_{ef_avg} to be less than 0, whereas v_{b_avg} remains unchanged. When an OCSF occurs on S_{12}, during postfault operation, when the current conduction path changed instead of flowing through the S_{12} under normal condition, the amplitude of v_a will reduce, causes v_{a_avg} to be less than $V_1/2$, whereas v_{b_avg} and v_{ef} remains unchanged. Remaining switches following the similar approaches.

According to the analysis, the characteristics of the average value after an OCSF occurs is summarized in Table 1, and the fault diagnose strategy can be derived from Table 1. The block diagram of the proposed diagnose strategy is shown in Fig. 9. By comparing the value with standard value, the faulty switch can be identified. The threshold voltage α is needed to set for nonideal factors, and the value of α is dependent on the converter hardware [7].

Table 1. The Characteristics of The Average Value after an OCSF Occurs

Switch	$v_{a\ avg}$	$v_{b\ avg}$	$v_{ef\ avg}$	Switch	$v_{c\ avg}$	$v_{d\ avg}$	$v_{gh\ avg}$
S_{11}	$<V_1/2$	$V_1/2$	<0	S_{21}	$<V_2/2$	$V_2/2$	<0
S_{12}	$<V_1/2$	$V_1/2$	0	S_{22}	$<V_2/2$	$V_2/2$	0
S_{13}	$>V_1/2$	$V_1/2$	0	S_{23}	$>V_2/2$	$V_2/2$	0
S_{14}	$>V_1/2$	$V_1/2$	>0	S_{24}	$>V_2/2$	$V_2/2$	>0
S_{15}	$V_1/2$	$<V_1/2$	>0	S_{25}	$V_2/2$	$<V_2/2$	>0
S_{16}	$V_1/2$	$<V_1/2$	0	S_{26}	$V_2/2$	$<V_2/2$	0
S_{17}	$V_1/2$	$>V_1/2$	0	S_{27}	$V_2/2$	$>V_2/2$	0
S_{18}	$V_1/2$	$>V_1/2$	<0	S_{28}	$V_2/2$	$>V_2/2$	<0

979-8-3315-2490-6/25 $31.00 © 2025 IEEE

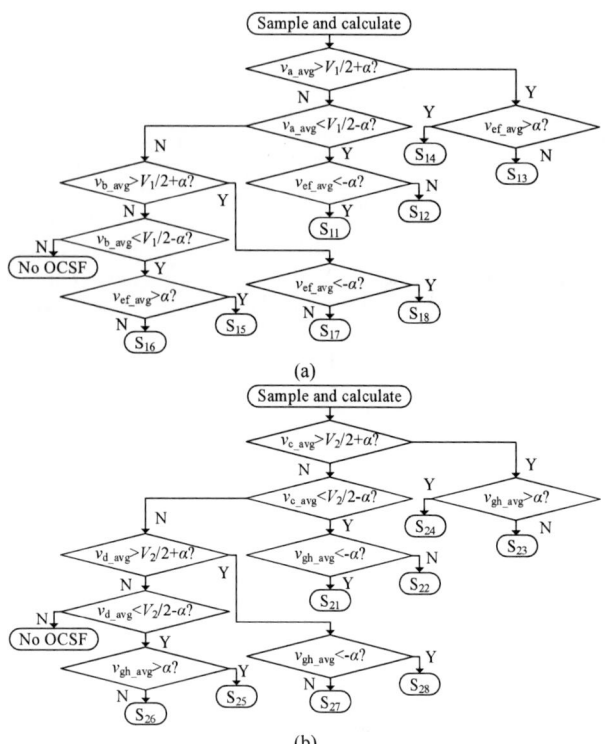

(a)

(b)

Fig. 9. Block Diagram of The OCSF Diagnose Strategy. (a) Block Diagram for The Primary Side Bridge, (b) Block Diagram for The Secondary Side Bridge.

IV. FAULT-TOLERANT STRATEGY OF FC-DAB

After the faulty switch is identified, in order to mitigate the negative effects caused by OCSF, the fault-tolerant strategy is proposed.

A. Primary Side Bridge Switch Fault

When an OCSF occurs on the primary side bridge, in order to restore the symmetry operational characteristics of the converter and maintain the voltage balance of the flying capacitors, the driving signals of the complementary switch are recommended to be blocked. The switches on primary side bridge are divided into four groups: S_{11} and S_{13}, S_{12} and S_{14}, S_{15} and S_{17}, S_{16} and S_{18}. When the faulty switch is identified, the complementary switch to be blocked can be selected from the corresponding group.

Only the condition $1<k<2$ and $(2-k)/4<d<0.5$ is presented, with remaining conditions following the similar approaches. The steady state waveforms when an OCSF occurs on S_{11} using the fault-tolerant strategy are shown in Fig. 10. With the fault-tolerant strategy, the driving signals of S_{13} is blocked. In $[t_0, t_1)$, the current flows through D_{18}, D_{17}, D_{12} and D_{11}, as shown in Fig. 11(a). In this state, $v_{ab}=V_1$, and $nv_{cd}=-nV_2$. Because $V_1>-nV_2$, i_L keep increasing. Then, i_L changes from negative to positive at t_1. In $[t_1, t_2)$, due to the positive i_L cannot flow through the faulty switch S_{11}, the current flow through D_{14}, C_{f11}, S_{12}, S_{17} and S_{18}, as shown in Fig. 11(b). In this state, v_{ab} reduce from V_1 to $V_1/2$, and $nv_{cd}=-nV_2$. Because $V_1/2>-nV_2$, i_L keep increasing. In $[t_2, t_3)$, the current conduction path of the primary side bridge is the same as $[t_1, t_2)$. In this state, $v_{ab}=V_1/2$,

and $nv_{cd}=nV_2$. Because $V_1/2<nV_2$, i_L keep decreasing. In $[t_3, t_4)$, the current flows through D_{14}, D_{13}, D_{16} and D_{15}, as shown in Fig. 11(c). In this state, $v_{ab}=-V_1$, and $nv_{cd}=nV_2$. Because $-V_1<nV_2$, i_L keep decreasing. Then, i_L changes from positive to negative at t_4. In $[t_4, t_5)$, due to the negative i_L cannot flow through the blocked switch S_{13}, the current flow through S_{15}, S_{16}, D_{12}, C_{f11} and S_{14}, as shown in Fig. 11(d). In this state, v_{ab} increase from $-V_1$ to $-V_1/2$, and $nv_{cd}=nV_2$. Because $-V_1/2<nV_2$, i_L keep decreasing. In $[t_5, t_6)$, the current conduction path of the primary side bridge is the same as $[t_4, t_5)$. In this state, $v_{ab}=-V_1/2$, and $nv_{cd}=-nV_2$. Because $-V_1/2>-nV_2$, i_L keep increasing. With complementary working states, symmetry operational characteristics is restored. The current values in $t_0\sim t_5$ can be calculated as follows.

$$
\begin{cases}
i_L(t_0) = \dfrac{T_{hs}nV_2}{L}\dfrac{(k+1)(2-k-4d)}{3k+4} \\[2mm]
i_L(t_1) = 0 \\[2mm]
i_L(t_2) = \dfrac{T_{hs}nV_2}{2L}\dfrac{(k+2)(3dk-k+2)}{3k+4} \\[2mm]
i_L(t_3) = \dfrac{T_{hs}nV_2}{L}\dfrac{(k+1)(4d+k-2)}{3k+4} \\[2mm]
i_L(t_4) = 0 \\[2mm]
i_L(t_5) = \dfrac{T_{hs}nV_2}{2L}\dfrac{(k+2)(k-3dk-2)}{3k+4}
\end{cases}
\tag{5}
$$

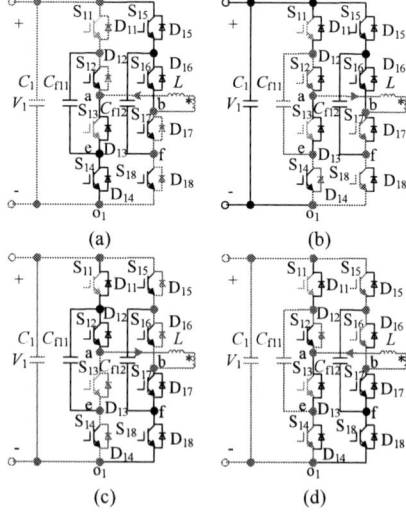

Fig. 10. Steady State Waveforms with The Fault-Tolerant Strategy When an OCSF Occurs on S_{11}.

(a) (b)

(c) (d)

Fig. 11. Current Conduction Path of The Primary Side Bridge with The Fault-Tolerant Strategy When an OCSF Occurs on S_{11}. (a) In $[t_0, t_1)$, (b) In $[t_1, t_2)$, (c) In $[t_3, t_4)$, (d) In $[t_4, t_5)$.

The unified transmission power P^* can be calculated as:

979-8-3315-2490-6/25 $31.00 © 2025 IEEE

$$P^* = \frac{1}{(3k+4)^2}(12dk^2 - 18d^2k^2 - 72d^2k - \tag{6}$$

$$48d^2 + 72dk + 48d + 4k^2 - 6k - 4)$$

The transmission power with the fault-tolerant strategy on the primary side bridge is related to k and d. The maximum unified transmission power can be expressed as:

$$P_{max}^* = \frac{2k^2 + 6k + 4}{3k^2 + 12k + 8} \tag{7}$$

Remaining switches on the primary side bridge following the similar approaches.

B. Secondary Side Bridge Switch Fault

Similarly, when an OCSF occurs on the secondary side bridge, the driving signals of the complementary switch are recommended to be blocked. The switches on secondary side bridge are divided into four groups: S_{21} and S_{23}, S_{22} and S_{24}, S_{25} and S_{27}, S_{26} and S_{28}. When the faulty switch is identified, the complementary switch to be blocked can be selected from the corresponding group.

Only the condition $k>1$ and $(k-1)/2k<d<0.5$ is presented, with remaining conditions following the similar approaches. The steady state waveforms when an OCSF occurs on S_{21} using the fault-tolerant strategy are shown in Fig. 12. With the fault-tolerant strategy, the driving signals of S_{23} is blocked. In $[t_0, t_1)$, the current flows through D_{28}, D_{27}, D_{22} and D_{21}, as shown in Fig. 13(a). In this state, $v_{ab}=V_1$, and $nv_{cd}=nV_2$. Because $V_1>nV_2$, i_L keep increasing. In $[t_1, t_2)$, the current conduction path of the secondary side bridge is the same as $[t_0, t_1)$. In this state, $v_{ab}=-V_1$, and $nv_{cd}=nV_2$. Because $-V_1<nV_2$, i_L keep decreasing. Then, i_L changes from positive to negative at t_2. In $[t_2, t_3)$, due to the negative i_L cannot flow through the faulty switch S_{21}, the current flow through D_{24}, C_{f21}, S_{22}, S_{27} and S_{28}, as shown in Fig. 13(b). In this state, nv_{cd} reduce from nV_2 to $nV_2/2$, and $v_{ab}=-V_1$. Because $-V_1<nV_2/2$, i_L keep decreasing. In $[t_3, t_4)$, the current flows through D_{24}, D_{23}, D_{26} and D_{25}, as shown in Fig. 13(c). In this state, $v_{ab}=-V_1$, and $nv_{cd}=-nV_2$. Because $-V_1<-nV_2$, i_L keep decreasing. In $[t_4, t_5)$, the current conduction path of the secondary side bridge is the same as $[t_3, t_4)$. In this state, $v_{ab}=V_1$, and $nv_{cd}=-nV_2$. Because $V_1>-nV_2$, i_L keep increasing. Then, i_L changes from negative to positive at t_5. In $[t_5, t_6)$, due to the positive i_L cannot flow through the blocked switch S_{23}, the current flow through S_{25}, S_{26}, D_{22}, C_{f21} and S_{24}, as shown in Fig. 13(d). In this state, nv_{cd} increase from $-nV_2$ to $-nV_2/2$, and $v_{ab}=V_1$. Because $V_1>-nV_2/2$, i_L keep increasing. With complementary working states, symmetry operational characteristics is restored. The current values in $t_0\sim t_5$ can be calculated as follows.

$$\begin{cases} i_L(t_0) = \dfrac{T_{hs}nV_2}{L}\dfrac{(k-1)(1-2k-3d)}{4k-3} \\[2mm] i_L(t_1) = \dfrac{T_{hs}nV_2}{L}\dfrac{(2k-1)(2dk-k+1)}{4k-3} \\[2mm] i_L(t_2) = 0 \\[2mm] i_L(t_3) = \dfrac{T_{hs}nV_2}{L}\dfrac{(k-1)(2k-1+3d)}{4k-3} \\[2mm] i_L(t_4) = \dfrac{T_{hs}nV_2}{L}\dfrac{(2k-1)(k-2dk-1)}{4k-3} \\[2mm] i_L(t_5) = 0 \end{cases} \tag{8}$$

Fig. 12. Steady State Waveforms with The Fault-Tolerant Strategy When an OCSF Occurs on S_{21}.

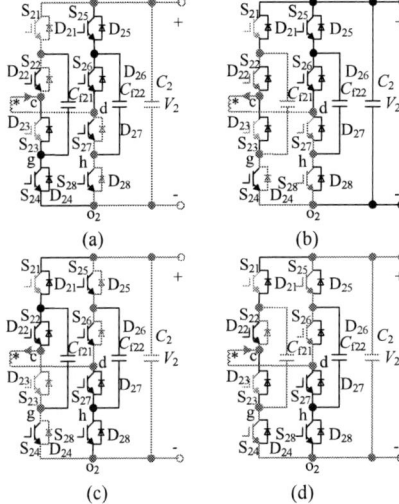

Fig. 13. Current Conduction Path of The Secondary Side Bridge with The Fault-Tolerant Strategy When an OCSF Occurs on S_{21}. (a) In $[t_0, t_1)$, (b) In $[t_2, t_3)$, (c) In $[t_3, t_4)$, (d) In $[t_5, t_6)$.

The unified transmission power P^* can be calculated as:

$$P^* = \frac{1}{(4k+3)^2}(48dk^2 - 48d^2k^2 - 72d^2k - \tag{6}$$

$$36d^2 + 72dk + 48d + 4k^2 + 6k - 10)$$

The transmission power with the fault-tolerant strategy on the secondary side bridge is related to k and d, similarly. The maximum unified transmission power can be expressed as:

$$P_{max}^* = \frac{16k^2 + 24k + 5}{(4k+3)^2} \tag{7}$$

Remaining switches on the secondary side bridge following the similar approaches.

V. SIMULATION RESULTS

In order to verify the effectiveness of the proposed fault diagnosis and fault-tolerant control strategy, simulation tests base on the topological structure of FC-DAB converter are performed. The main parameters of the FC-DAB converter for the simulation tests are summarized in Table 2.

Table 2. The Main Parameters for The Simulation Tests

Parameter	Value	Parameter	Value
V_1	48 V	L	60 μH
V_2	40 V	C_{f11}~C_{f18}, C_{f21}~C_{f28}	470 μF
f_s	10 kHz	C_1, C_2	470 μF
n	1	d	0.3

The simulated waveforms when an OCSF occurs on S_{11} is shown in Fig. 14(a). In Fig. 14(a), Fault occurs at the time t=10 μs, corresponding to the rising edge of the driving signal for S_{11} in normal operation state. From Fig. 14(a), compared with normal operation, during postfault operation, the amplitude of v_a and v_{ef} are reduced, whereas v_b remains unchanged. And with proposed fault diagnose strategy, the faulty switch S_{11} is identified effectively. The simulated waveforms when an OCSF occurs on S_{12} is shown in Fig. 14(b). In Fig. 14(b), Fault occurs at the time t=10 μs, corresponding to the rising edge of the driving signal for S_{12} in normal operation state. From Fig. 14(b), compared with normal operation, during postfault operation, the amplitude of v_a is reduced, whereas v_b and v_{ef} remain unchanged. And with proposed fault diagnose strategy, the faulty switch S_{12} is identified effectively.

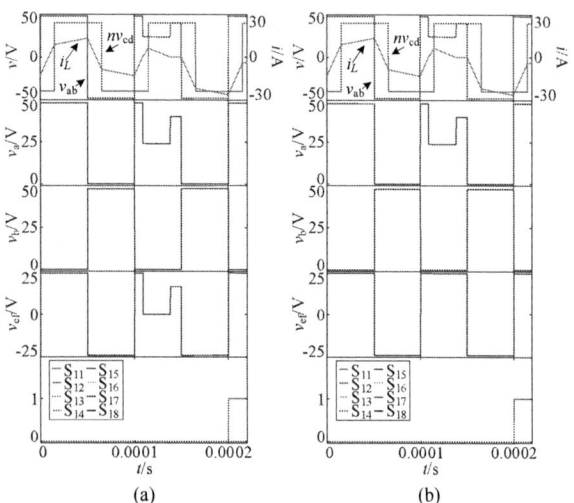

Fig. 14. Simulation Waveforms when an OCSF Occurs on The Primary Side Bridge. (a) OCSF Occurs on S_{11}, (b) OCSF Occurs on S_{12}.

The simulated waveforms after using the fault-tolerant control strategy when an OCSF occurs on S_{11} and S_{12} are shown in Fig. 15(a) and Fig. 15(b), respectively. From Fig. 15, the symmetry operational characteristics of the FC-DAB converter is restored.

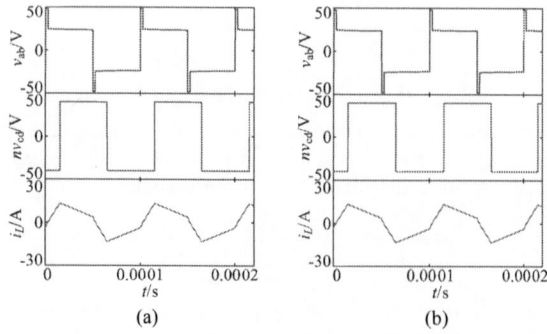

Fig. 15. Simulation Waveforms after Using The Fault-Tolerant Control Strategy when an OCSF Occurs on The Primary Side Bridge. (a) OCSF Occurs on S_{11}, (b) OCSF Occurs on S_{12}.

The simulated waveforms when an OCSF occurs on S_{21} is shown in Fig. 16(a). In Fig. 16(a), Fault occurs at the time t=10 μs, corresponding to the rising edge of the driving signal for S_{21} in normal operation state. From Fig. 16(a), compared with normal operation, during postfault operation, the amplitude of v_c and v_{gh} are reduced, whereas v_d remains unchanged. And with proposed fault diagnose strategy, the faulty switch S_{21} is identified effectively. The simulated waveforms when an OCSF occurs on S_{22} is shown in Fig. 16(b). In Fig. 16(b), Fault occurs at the time t=10 μs, corresponding to the rising edge of the driving signal for S_{22} in normal operation state. From Fig. 16(b), compared with normal operation, during postfault operation, the amplitude of v_c is reduced, whereas v_d and v_{gh} remain unchanged. And with proposed fault diagnose strategy, the faulty switch S_{22} is identified effectively.

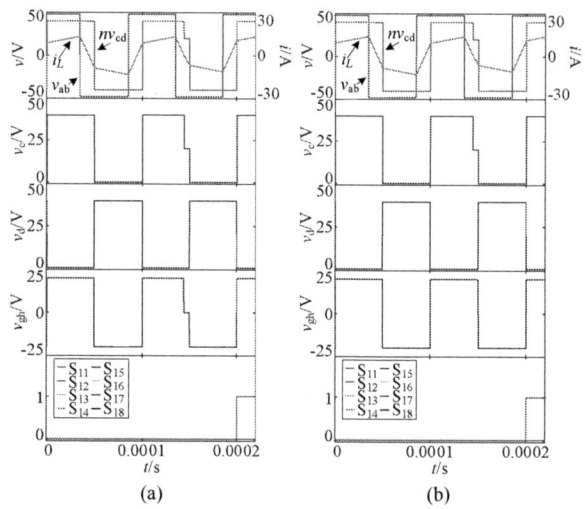

Fig. 16. Simulation Waveforms when an OCSF Occurs on The Secondary Side Bridge. (a) OCSF Occurs on S_{21}, (b) OCSF Occurs on S_{22}.

The simulated waveforms after using the fault-tolerant control strategy when an OCSF occurs on S_{21} and S_{22} are shown in Fig. 17(a) and Fig. 17(b), respectively. From Fig. 17, the symmetry operational characteristics of the FC-DAB converter is restored.

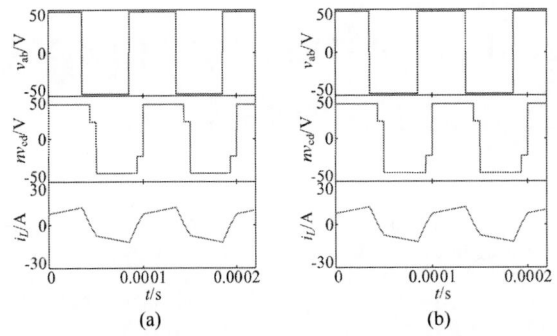

(a)   (b)

Fig. 17. Simulation Waveforms after Using The Fault-Tolerant Control Strategy when an OCSF Occurs on The Secondary Side Bridge. (a) OCSF Occurs on S_{21}, (b) OCSF Occurs on S_{22}.

VI. CONCLUSION

This paper proposes an open-circuit switch fault diagnosis and fault-tolerant control strategy for the FC-DAB converter. The proposed fault diagnose strategy utilizes the average values of the midpoint voltages of the bridge arms and the low-potential side voltages of flying capacitors to identify the faulty switch. The proposed fault-tolerant control strategy blocks the complementary switch to restore symmetry operational characteristics after the faulty switch is identified, thus mitigates the negative effects caused by OCSF and enhances the operational reliability of the FC-DAB converter. Simulation results show that the faulty switch can be identified effectively with the proposed fault diagnose strategy, and symmetry of operational characteristics is restored with the proposed fault-tolerant control strategy. This study is based on idealized model parameters, and future research will consider the impact of parasitic parameters.

REFERENCES

[1] Li L T, Xu G, Sha D S, et al. Review of dual-active-bridge converters with topological modifications[J]. IEEE Transactions on Power Electronics, 2023, 38(7): 9046-9076.

[2] Higa H, Itoh J. Derivation of operation mode for flying capacitor topology applied to three-level DAB converter[C]//2015 IEEE 2nd International Future Energy Electronics Conference (IFEEC), 2015: 1-6.

[3] Wen H Q, Li J L, Shi H C, et al. Fault diagnosis and tolerant control of dual-active-bridge converter with triple-phase-shift control for bidirectional EV charging systems[J]. IEEE Transactions on Transportation Electrification, 2021, 7(1): 287-303.

[4] Davoodi A, Noroozi N, Zolghadri M R. A fault-tolerant strategy for three-phase dual active bridge converter[C]//2019 10th International Power Electronics, Drive Systems and Technologies Conference (PEDSTC), 2019: 253-258.

[5] Airabella A M, Oggier G G, Piris-Botalla L E, et al. Semi-conductors faults analysis in dual active bridge DC–DC converter[J]. IET Power Electronics, 2016, 9(6): 1103-1110.

[6] Zheng M K, Wen H Q, Shi H C, et al. Open-circuit fault diagnosis of dual active bridge DC-DC converter with extended-phase shift control[J]. IEEE Access, 2019, 7: 23752-23765.

[7] Song C C, Sangwongwanich A, Yang Y H, et al. Open-circuit fault diagnosis and tolerant control for 2/3-Level DAB converters[J]. IEEE Transactions on Power Electronics, 2023, 38(4): 5392-5410.

[8] Zhao N, Liu J Q, Shi Y M, et al. mode analysis and fault-tolerant method of open-circuit fault for a dual active-bridge dc–dc converter[J]. IEEE Transactions on Industrial Electronics, 2020, 67(8): 6916-6926.

[9] Wu J J, Wang H Y, Ma M Y, et al. Mode analysis and fault-tolerant method of open-circuit fault for a three-level dual active bridge DC-DC converter[J]. Microelectronics Reliability, 2023, 150: 115100.

Research on Cooperative Operation Mode of Pumped Storage Cluster for Regional Power Grid

Jiaxing Wang
State Key Laboratory of
Alternate Electrical Power
System with Renewable Energy
Sources
North China Electric Power
University
Beijing, China
ncepu_wjx@163.com

Chao Huo
Power Grid Technical Support
Center
Northwest Branch of State Grid
Corporation of China
Xi'an, China
huochao001@126.com

Yuxuan Yang
State Key Laboratory of
Alternate Electrical Power
System with Renewable Energy
Sources
North China Electric Power
University
Beijing, China
yixian_yang@163.com

Xiuting Rong
Power Grid Technical Support
Center
Northwest Branch of State Grid
Corporation of China
Xi'an, China
rongxiuting@126.com

Zun Guo
Transmission Power Grid
Planning Center
State Grid Economic and
Technological Research Institute
Company Ltd.
Beijing, China
zunguo93ee@163.com

Zhi An
Transmission Power Grid
Planning Center
State Grid Economic and
Technological Research Institute
Company Ltd.
Beijing, China
thuaz07@gmail.com

Ming Zhou*
State Key Laboratory of
Alternate Electrical Power
System with Renewable Energy
Sources
North China Electric Power
University
Beijing, China
zhouming@ncepu.edu.cn
*Corresponding author

Zhaoyuan Wu
State Key Laboratory of
Alternate Electrical Power
System with Renewable Energy
Sources
North China Electric Power
University
Beijing, China
wuzy@ncepu.edu.cn

Abstract—**As the penetration of renewable energy continues to rise, the power system's flexible resources are increasingly strained, posing significant challenges to new energy consumption and the secure, stable operation of the grid. As a large-scale energy storage and balancing resource, pumped storage hydropower (PSH) can effectively mitigate the impact of renewable energy integration on the power system. Supported by favorable policies and abundant geographical resources, China is poised to develop multiple pumped storage clusters (PSCs) in the coming years. To investigate the coordinated operation of PSC with other energy resources in regional grid, this study first defines PSC and constructs a corresponding model. Subsequently, an optimized dispatching model is developed to minimize total operational costs by integrating PSC with wind, solar, hydro, and thermal power systems. A case study is conducted using a representative region in China, considering both typical and extreme weather scenarios, to evaluate the influence of the configuration of PSC on the system under different scenarios. The results indicate that the integration of PSC can significantly enhance the new energy consumption and reduce operational costs, which demonstrates the effectiveness of the proposed model and the crucial role of PSC in promoting low-carbon, reliable, and economically efficient operation of power system operation.**

Keywords—*pumped storage cluster (PSC), economic dispatch model, coordinated operation mode, evaluation of operational effects*

I. INTRODUCTION

To achieve the "dual carbon" target, China is accelerating the development of a clean, low-carbon, safe, and efficient energy system. The installed capacity of new energy sources—primarily wind and solar power—continues to expand within the power system [1]. As of the end of March 2025, China's total installed power generation capacity reached approximately 3.43 billion kilowatts, of which the installed capacity of new energy is about 1.49 billion kilowatts, accounting for 43.4 % of the total installed capacity. Due to the uncertainty and intermittency of photovoltaic output, as well as the anti-peak-shaving characteristics of wind power generation, the large-scale integration of these resources into the grid poses significant challenges to new energy consumption and the safe and stable operation of the power system [2]. Energy storage has the advantages of fast response and flexible regulation. It is not only an essential component of the new type power system but also a key support for the emerging energy framework. Rational allocation of energy storage resources is an effective means to promote the integration and consumption of new energy, mitigate power output fluctuations, and enhance the operational flexibility of the power system [3]. Considering the techno-economic characteristics of various types of energy storage and regulation resources, pumped storage hydropower (PSH) stands out as the most technically mature and economically viable option for large-scale energy storage and grid regulation. It also offers unique advantages in terms of deployment scale and service life [4]. In recent years, driven by China's abundant pumped storage site resources, the growing demand for flexible regulation in the power system, and supportive policies, PSH has experienced rapid development. In the future, multiple pumped storage clusters (PSCs) are expected to emerge in China, forming a synergistic relationship with the development and utilization of renewable energy. Therefore, against the backdrop of an increasingly diversified power generation mix, studying the coordinated operation mode between PSC and other resources within the regional power grid is of great significance for reducing renewable energy curtailment and enhancing system flexibility.

Project Supported by the Science and Technology Project of State Grid
Corporation of China (5100-202306427A-3-2-ZN).

At present, there is limited research on the coordinated operation of PSC with other resources within the regional power grid. Most studies by domestic and international scholars focus on the complementary operation between individual pumped storage station and renewable energy sources or emerging energy storage. P. Xia et al. [5] proposed a dispatching model for optimizing the operation of the PSH in the system to solve the problem of difficulty in wind power consumption. Y. Zhou et al. [6] developed a joint optimization dispatching model for renewable energy and pumped storage hydropower to improve power supply reliability and facilitate large-scale renewable integration. X. Liu et al. [7] introduced a model to explore typical peak shaving strategies for cascaded hybrid pumped storage systems. A.A. Bafrani et al. [8] accounted for the stochastic fluctuations of renewable output and applied a robust optimization approach to develop a joint operation model for renewables and pumped storage. J. Tan et al. [9] presented an optimization model that coordinates cascaded hydropower and pumped storage, considering hydro-solar-pumped storage complementarity and DC power transmission. M. S. Javed et al. [10] proposed an optimal dispatch strategy for combined pumped storage and battery energy storage systems to enhance the reliability of off-grid renewable energy systems. W. Huang et al. [11] constructed a joint scheduling model of wind-photovoltaic-electrochemical energy storage-pumping and storage to study the operation characteristics and complementary characteristics of pumping and electrochemical energy storage.

In summary, although existing studies have explored the complementary operation between PSH and renewable energy, the capacity constraints of PSH are mostly represented in terms of energy limits, which fails to accurately reflect the relationship between reservoir storage and energy in practical operations. Moreover, most existing research focuses on individual PSH within the regional power grid, with an emphasis on its roles in smoothing renewable energy fluctuations and providing peak shaving. However, studies exploring the role of PSC as flexible regulating resource for peak demand assurance and renewable energy integration remain scarce, making current research insufficiently aligned with China's future development needs. In addition, very few studies have taken into account the effect of pumped storage clusters on the power grid in extreme weather scenario. Therefore, this study first introduces the concept of a PSC and establishes mathematical models for various types of PSH within the cluster. Then, a day-ahead coordinated dispatching model is developed to optimize the operation of the PSC and other regional grid resources, with the objective of minimizing system operating cost. Then, using a representative region in China as a case study, typical and extreme weather scenarios are constructed by accounting for both load characteristics and renewable generation profiles. A comparative analysis is conducted between the PSC and traditional cascade hydropower stations under both scenarios. The results demonstrate the advantages of PSC in enhancing renewable energy utilization, reducing system operating costs, and substituting thermal power generation.

II. PUMPED STORAGE CLUSTER ANALYSIS AND MODELING

The PSC proposed in this study refers to a coordinated multi-station system, which consists of multiple PSHs located within a single river basin. These plants form an integrated system, which is jointly managed to enhance overall operational efficiency. The PSC aims to maximize the utilization of limited water resources through coordinated control and dispatch. Through collaborative reservoir scheduling, the PSC addresses the demand for flexible regulation in new power system. Currently, three main types of PSHs have been built or planned in China.

A. Type of Pumped Storage Hydropower

1) Pure Pumped Storage Hydropower

The pure pumped storage hydropower (PPSH) refers to the power station where the upper reservoir has little or no natural inflow, and all the water required for generating or pumping operations is cycled between the upper and lower reservoirs. The station is primarily responsible for peak shaving, valley filling, and providing emergency backup, rather than conventional power generation tasks. Due to its relatively flexible site selection, PPSH is typically located near load center to reduce power transmission loss and minimize investment in transmission lines.

2) Hybrid Pumped Storage Hydropower

The hybrid pumped storage hydropower (HPSH) refers to the power station with natural runoff inflow in the upper reservoir, including hydropower units and reversible units or pumping stations. This type of plant can utilize existing conventional hydropower units to perform regular hydropower generation and comprehensive water resource utilization, while also supporting peak shaving, valley filling, and emergency backup functions. The HPSH is mostly developed by retrofitting traditional hydropower stations with pumped storage units or by adding pumping stations.

3) Shared Reservoir Pumped Storage Hydropower

The shared reservoir pumped storage hydropower (SRPSH) means that the upper reservoir of the power station is built on the top platform of the mountain, and the lower reservoir shares the reservoir of the hydropower station. The land area and cost required for the construction of this type of power station are low, but the site selection requirements are more stringent.

B. Pumped Storage Cluster

The PSC developed in this study includes two types of stations: HPSH and SRPSH. Specifically, the cluster is based on four cascaded hydropower stations, where pumped storage units are added to the first and second level stations to form HPSHs, while the reservoir of the fourth level station is shared with a dedicated PSH. The PSC model incorporates both internal constraints of pumped storage units and overall station-level constraints:

1) Pumped Storage Unit Constraints

The constraints on pumped storage units include operating status, power output, water flow, minimum continuous startup/shutdown time, and start-stop times. These aspects have been discussed in detail in [11-12] and are therefore not elaborated further here.

2) Pumped Storage Cluster Constraints

a) Water Balance Constraint

The first, second and third level power stations:

$$V_{p,t} = V_{p,t-1} + \left(\begin{array}{c} \sum\limits_{i=1}^{I}(u_{p-1,i,t}^{G}Q_{p-1,i,t}^{G} - u_{p-1,i,t}^{P}Q_{p-1,i,t}^{P}) + \\ \sum\limits_{h=1}^{H}u_{p-1,h,t}Q_{p-1,h,t} + q_{p-1,t} + \\ \sum\limits_{i'=1}^{I'}(u_{p,i',t}^{P}Q_{p,i',t}^{P} - u_{p,i',t}^{G}Q_{p,i',t}^{G}) - \\ \sum\limits_{h'=1}^{H'}u_{p,h',t}Q_{p,h',t} - q_{p,t} + I_{1,t} \end{array} \right) \Delta t \quad (1)$$

Where $V_{p,t}$ denotes the reservoir storage of station p at time period t; $u_{p,i,t}$ is 0-1 variables of unit i of station p operating in time period t; $Q_{p,i,t}$ represent the generating flow; I_t and $q_{p,t}$ represent the reservoir inflow (considered only for the first-level station) and water spillage, respectively; $u_{p,h,t}$ and $Q_{p,h,t}$ denote the operational status and generation flow of hydropower unit h at time t;H is the total number of hydropower units at station p. Unless otherwise specified, all variables with superscripts G and P in this study represent corresponding values under generation and pumping conditions, respectively.

The fourth level power station:

$$V_{p,t} = V_{p,t-1} + \left(\begin{array}{c} \sum\limits_{ip=1}^{IP}(u_{ip,t}^{G}Q_{ip,t}^{G} - u_{ip,t}^{P}Q_{ip,t}^{P}) + \sum\limits_{h=1}^{H}u_{p-1,h,t}Q_{p-1,h,t} + \\ q_{p-1,t} - \sum\limits_{h'=1}^{H'}u_{p,h',t}Q_{p,h',t} - q_{p,t} \end{array} \right) \Delta t \quad (2)$$

Where IP is the total number of pumped storage units in the SRPSH.

b) Reservoir Capacity and Water Level Constraints

$$V_p^{\min} \le V_{p,t} \le V_p^{\max} \quad (3)$$

$$\begin{cases} H_p^{F,\min} \le H_{p,t}^{F} \le H_p^{F,\max} \\ H_{p,0}^{F} = H_{p,end}^{F} \\ H_{p,t} = \dfrac{(H_{p,t}^{F} + H_{p,t-1}^{F})}{2} - H_{p,t}^{T} \end{cases} \quad (4)$$

Where V_p^{min} and V_p^{max} represent the upper and lower bounds of the reservoir storage capacity of station p; $H_{p,t}^{F}$ and $H_{p,t}^{T}$ denote the average forebay and tailwater levels of station p during time period t, respectively; $H_p^{F,min}$ and $H_p^{F,max}$ represent the dead water level and normal storage level of the forebay of station p, respectively; $H_{p,0}^{F}$ and $H_{p,end}^{F}$ are the forebay levels at the beginning and end of the dispatching period, respectively; $H_{p,t}$ denotes the net head of station p during power generation.

c) Forebay Level–Reservoir Volume and Tailwater Level–Discharge Flow Constraints

$$\begin{cases} H_{p,t}^{F} = f_{hv}^{F}(V_{p,t}) \\ H_{p,t}^{T} = f_{hq}^{T}(Q_{p,t}) \end{cases} \quad (5)$$

Where $f_{hv}^{F}()$ is the function describing the relationship between the forebay water level and the reservoir volume of the station; $f_{hv}^{T}()$ is the function representing the relationship between the tailwater level and the discharge flow of the station; Q_t denotes the discharge flow of the station at time period t, considering only generation discharge and water spillage in this study.

III. COORDINATED OPTIMIZATION DISPATCHING MODEL OF PUMPED STORAGE CLUSTER AND OTHER RESOURCES

The regional power grid considered in this study includes wind power, solar power, thermal power, and hydropower resources. By integrating PSC, the regulation capability of hydropower within the basin can be enhanced, thereby improving resource allocation in the regional grid, and facilitating new energy consumption, and ultimately enhancing the overall economic efficiency of renewable energy development [13]. Considering economic and environmental objectives, and incorporating constraints such as system power balance and the operational limits of generating units, this study develops a day-ahead dispatching optimization model that aims to minimize the total system operating cost.

A. Objective Function

$$\min f = \sum_{t=1}^{T}\left(C_{gen,t} + C_{aba,t} + C_{curt,t} + C_{c,t} \right) \quad (6)$$

Where f denotes the total system operating cost; T is the total number of dispatching periods; $C_{gen,t}$, $C_{aba,t}$, $C_{curt,t}$ and $C_{c,t}$ represent the thermal unit operating cost, new energy curtailment penalty cost, load loss penalty cost, and carbon emission cost during time period t, respectively.

B. Operation Constraints

1) Node Power Balance Constraint

$$\sum_{g\in G(j)} P_{g,t} + \sum_{h\in H(j)} P_{h,t} + \sum_{w\in W(j)} P_{w,t} + \sum_{s\in S(j)} P_{s,t} +$$
$$\sum_{p\in P(j)}\left(P_{p,t}^{G} - P_{p,t}^{P} \right) = \sum_{d\in D(j)}\left(P_{d,t} - P_{d,t}^{cul} \right) \quad (7)$$

Where $P_{g,t}$, $P_{h,t}$, $P_{w,t}$, and $P_{s,t}$ represent the output power of thermal units, hydropower units, wind farms, and photovoltaic stations at time period t, respectively; $P_{p,t}^{G}$ and $P_{p,t}^{P}$ denote the generation and pumping power of PSHs, respectively; $P_{d,t}$ and $P_{d,t}^{cul}$ represent the system load and the amount of load loss, respectively; G, H, W, S, P, and D denote the sets of nodes associated with thermal units, hydropower units, wind farms, photovoltaic stations, PSHs, and loads, respectively. *Wind Power Operation Constraint*

$$0 \le P_{w,t} \le P_{w,t}^{pre} \quad (8)$$

Where $P_{w,t}^{pre}$ is the predicted value of wind power output in t period.

3) Wind Power Operation Constraint

$$0 \le P_{s,t} \le P_{s,t}^{pre} \tag{9}$$

Where $P_{s,t}^{pre}$ is the predicted value of wind power output in t period.

4) Thermal Power Operation Constraints

a) Power Output Constraint

$$x_{g,t} P_g^{min} \le P_{g,t} \le x_{g,t} P_g^{max} \tag{10}$$

Where $x_{g,t}$ is the operating variable; P_g^{min} and P_g^{max} are the minimum output and maximum output, respectively.

b) Ramp Rate Constraint

$$\begin{cases} P_{g,t} - P_{g,t-1} \le P_g^{min}\left(x_{g,t} - x_{g,t-1}\right) + P_g^{max}\left(1 - x_{g,t}\right) + R_g^u x_{g,t-1} \\ P_{g,t-1} - P_{g,t} \le P_g^{min}\left(x_{g,t-1} - x_{g,t}\right) + P_g^{max}\left(1 - x_{g,t-1}\right) + R_g^d x_{g,t} \end{cases} \tag{11}$$

Where R_g^u and R_g^d are the extreme values of upward and downward climbing of thermal power units at time period t, respectively.

5) Hydropower Operation Constraints
The constraints are detailed in [11-12].

6) Pumped Storage Cluster Constraints
The constraints are detailed in Section II.

IV. CASE STUDY

To validate the feasibility and accuracy of the proposed model, a case study is conducted based on a regional power grid in China. Typical operational and extreme weather scenarios are constructed to investigate the coordinated operation of PSC with other resources within the regional grid. The power source configuration and load forecast are based on the region's 2030 planning data. The installed capacities are as follows: wind power 16,300 MW, solar power 49,000 MW, thermal power 8,590.5 MW, hydropower across four cascade stations 6,640 MW, and PSH 2,400 MW. Two retrofit schemes for the four-level hydropower system are explored in this study.

Scheme 1: Expand conventional hydropower units, increasing the total hydropower capacity to 9,080 MW.

Scheme 2: Retrofit the first and second-level hydropower stations by adding pumped storage units to form HPSHs. These, together with a PSH sharing the reservoir of the fourth-level station, constitute a PSC. The total installed capacity of the pumped storage units after retrofit reaches 4,840 MW.

A. Scenario Construction

The wind and PV output curves, as well as the load profile under the typical scenario, are constructed by scaling normalized historical representative curves according to the region's 2030 installed capacities. The normalized representative curves are derived using the K-means clustering method applied to historical data from a selected year, based on seasonal characteristics. This study selects a typical summer day for analysis. The extreme weather scenario is constructed based on realistic meteorological hazards, focusing on prolonged high-temperature and clear-sky event that may occur in summer. The data is constructed by modifying wind and solar power outputs, as well as load data, based on the typical scenario. The detailed scenario configuration is shown in TABLE I. The wind and solar power output curves and load curves under both the typical and extreme weather scenarios are illustrated in Fig. 1.

TABLE I. DETAILED SCENARIO CONFIGURATION

Scenario	Scenario description	Scenario construction
I	Typical summer day	Obtained by K-means clustering of normalized typical representative curves
II	High-temperature and clear-sky event	Wind power output decreases, solar output increases, and load increases

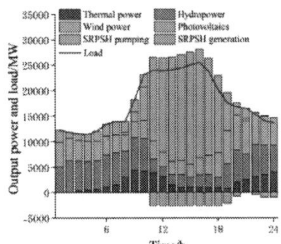

(a) Wind power   (b) Solar power   (c) Load

Fig. 1. Data curve diagram

B. Analysis of Dispatching Results

1) Dispatching Operation Results

Based on the coordinated optimization dispatch model of the PSC and other resources, the operational results under the typical scenario for the two retrofitting schemes are shown in Fig. 2. From 01:00 to 06:00, the objective function drives the system to minimize operating costs. As wind and hydropower incur no operational costs, they are prioritized for grid dispatch, while the remaining power deficit is compensated by the PSH. Between 06:00 and 10:00, system load increases while solar power output remains low. The combination of wind, hydropower, and PSH is insufficient to meet the demand, necessitating the startup of thermal power units. From 11:00 to 19:00, solar power output reaches a high level. To reduce both operational costs and curtailment, some thermal units are shut down, and the PSH switches to the pumping mode. Between 20:00 and 24:00, although system load decreases, wind power output remains high. Thus, the PSH continues in pumping mode to reduce renewable energy curtailment.

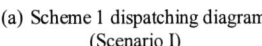

(a) Scheme 1 dispatching diagram (Scenario I)   (b) Scheme 2 dispatching diagram (Scenario I)

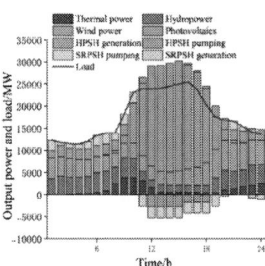

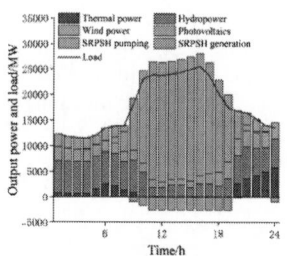 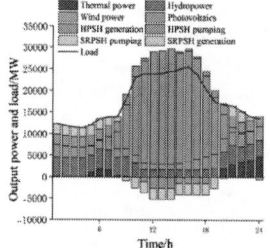

(c) Scheme 1 dispatching diagram   (d) Scheme 2 dispatching diagram
(Scenario II)   (Scenario II)

Fig. 2. System dispatching operation diagram

2) Analysis of System Operational Metrics

This study proposes a set of system performance evaluation metrics from the perspectives of economic efficiency, operational flexibility, and carbon reduction. Furthermore, the impacts of incorporating PSC on system operation are analyzed under both typical and extreme weather scenarios. The detailed results are presented in TABLE II.

TABLE II. SYSTEM OPERATION METRICS UNDER DIFFERENT SCENARIOS and SCHEMES

Scenario	Scheme	System operational Metrics				
		Economy ($)		Flexibility		Low carbon
		Thermal power operation cost	Curtailment penalty cost	New energy consumption	Thermal power start-stop count	Carbon emission (t)
I	1	1519128.32	615261.78	86.43%	40	35284.06
	2	979259.09	287303.05	93.66%	39	22556.75
II	1	1105200.25	524334.09	88.69%	42	25925.91
	2	765344.04	289044.03	93.76%	32	18479.69

1) From an economic perspective, PSC has the characteristics of large capacity and rapid capability. It can discharge power during peak demand periods and coordinate with thermal power to reduce thermal output, thereby reducing thermal operation costs. It can also store excess energy during periods of high wind and solar output, reducing curtailment and associated penalties. Under typical and extreme weather scenarios, thermal generation costs decrease by 35.5% and 30.8%, respectively, while curtailment penalty costs decline by 53.3% and 44.9%.

2) In terms of flexibility, the PSC enables real-time power adjustment and stores excess energy through pumping. Its integration significantly reduces wind and solar curtailment, thereby improving new energy utilization rate, which increases by 7.23% and 5.07% under the typical and extreme weather scenarios, respectively. Due to its rapid response capability, the PSC can partially replace thermal output during peak load periods, reducing reliance on thermal units. As a result, thermal power start-stop counts decrease by 1 instance under the typical scenario and by 10 instances under the extreme scenario, demonstrating a significant substitution capability.

3) From the perspective of carbon reduction, since carbon emissions mainly come from the operation of thermal power units, substituting thermal output with PSC leads to a significant reduction in system-wide carbon emissions.

V. CONCLUSION

This study begins by defining the concept of the PSC and establishing its corresponding mathematical model. To address the unclear positioning of PSC in regional power grids and the lack of clearly defined collaborative operation modes with other resources, a day-ahead joint optimization dispatch model is proposed. By comparing dispatch results with and without PSC under typical and extreme weather scenarios, the following conclusions are drawn:

1) With the integration of PSC, excess new energy can be stored via pumping during surplus periods, significantly improving the utilization rate of new energy. During periods of power shortage, it can provide timely output to meet demand, thus enhancing system reliability.

2) The integration of PSC effectively reduces the output and start-stop frequency of thermal units, leading to notable reductions in operational and start-up costs—especially under extreme weather scenario. Moreover, the use of PSC helps lower carbon emissions, contributing to energy conservation and emission reduction goals.

This paper evaluates system performance from the perspectives of economic efficiency, operational flexibility, and carbon reduction, without considering the impact of PSC on system security. Future research will focus on the role of PSC in frequency support and other aspects of grid stability.

REFERENCES

[1] Q. Huang, Y. Guo, J. Jiang, and B. Ming, "Development Pathway of China's Clean Electricity Under Carbon Peaking and Carbon Neutrality Goals," JOURNAL OF SHANGHAI JIAO TONG UNIVERSITY, vol. 55, no. 12, pp. 1499–1509, 2021.

[2] Y. An, S. Zheng, R. Su, and R. Yang, "Research On Two-Layer Optimization of Wind-Solar-Water-Storage Multi Energy Complementary Power Generation System," ACTA ENERGIAE SOLARIS SINICA, vol. 44, no. 12, pp. 510–517, 2023.

[3] X. Wang, and H. Zhong, "Research on the Synergetic Development of Pumped Storage and New Energy Storage under the Background of New Power System," Journal of Global Energy Interconnection, vol. 7, no. 4, pp. 363–371, 2024.

[4] K. R. Vasudevan, V. K. Ramachandaramurthy, G. Venugopal, J. B. Ekanayake, and S. K. Tiong, "Variable speed pumped hydro storage: A review of converters, controls and energy management strategies," Renewable and Sustainable Energy Reviews, vol. 135, p. 110156, Jan. 2021.

[5] P. Xia et al., "Robust Unit Commitment with Pumped Storage Units for Wind Power Accommodation," Automation of Electric Power Systems, vol. 42, no. 19, pp. 41–49, 2018.

[6] Y. Zhou et al., "Optimizing pumped-storage power station operation for boosting power grid absorbability to renewable energy," Energy Conversion and Management, vol. 299, p. 117827, Jan. 2024.

[7] X. Liu et al., "Short-term peak shaving model of cascade hybrid pumped storage retrofitted from conventional hydropower," Power System Technology, vol. 49, no. 3, pp. 1217–1226, 2025.

[8] A. A. Bafrani, A. Rezazade, and M. Sedighizadeh, "Robust electrical reserve and energy scheduling of power system considering hydro pumped storage units and renewable energy resources," Journal of Energy Storage, vol. 54, p. 105310, Oct. 2022.

[9] J. Tan et al., "Distributionally Robust Optimal Scheduling Method of Power System Considering Hydropower-Photovoltaic-Pumped Storage Complementarity and DC Transmission," Proceedings of the CSEE, vol. 44, no. 15, pp. 5947–5960, 2024.

[10] M. S. Javed, D. Zhong, T. Ma, A. Song, and S. Ahmed, "Hybrid pumped hydro and battery storage for renewable energy based power supply system," Applied Energy, vol. 257, p. 114026, Jan. 2020.

[11] W. Huang et al., "Multi-time scale joint optimal scheduling for wind-photovoltaic-electrochemical energy storage-pumped storage considering renewable energy uncertainty," Electric Power Automation Equipment, vol. 43, no. 4, pp. 91–98, 2023.

[12] H. Wang, S. Liao, C. Cheng, B. Liu, Z. Fang, and H. Wu, "Short-term scheduling strategies for hydro-wind-solar-storage considering variable-speed unit of pumped storage," Applied Energy, vol. 377, p. 124336, Jan. 2025.

[13] D. Han, W. Ren, L. Zhou,Z. Cui and G. He, "The Development Status and Outlook of Pumped Storage in China," Yangtze River, vol. 55, no. 11, pp. 39–45, 2024.

2025 International Conference on Advanced Energy Systems and Power Electronics (AESPE 2025)

Control of DFIG-Based Offshore Wind Farm with Hydrogen Production Connected to DRU-HVDC

Zheng Li*
Clean Energy Integration & Electrical Dept
Huaneng Clean Energy Research Institute
Beijing, China
z_li@qny.chng.com.cn

Junyang Zhang
Clean Energy Integration & Electrical Dept
Huaneng Clean Energy Research Institute
Beijing, China
jy_zhang2@qny.chng.com.cn

Cheng Peng
Clean Energy Integration & Electrical Dept
Huaneng Clean Energy Research Institute
Beijing, China
c_peng@qny.chng.com.cn

Yijing Chen
Clean Energy Integration & Electrical Dept
Huaneng Clean Energy Research Institute
Beijing, China
yj_chen@qny.chng.com.cn

Lei Ba
Clean Energy Integration & Electrical Dept
Huaneng Clean Energy Research Institute
Beijing, China
l_ba@qny.chng.com.cn

Chunhua Li
Clean Energy Integration & Electrical Dept
Huaneng Clean Energy Research Institute
Beijing, China
lchh_1990@126.com

Han Wu
Zhejiang University
College of Electrical Engineering
Hangzhou, China
han.wu@zju.edu.cn

Tao Wang
Zhejiang University
College of Electrical Engineering
Hangzhou, China
wangtaoee@zju.edu.cn

Abstract—**Offshore wind farms with hydrogen production can simultaneously provide hydrogen for hydrogen-powered ships and facilitate the consumption of offshore wind energy. With expansion of offshore wind into deeper-sea regions, platform cost and reliability have emerged as critical challenges. Hence, this paper proposes a scheme of the diode rectifier unit (DRU)-based offshore wind farm with hydrogen production. A grid-forming control strategy for doubly-fed induction generators (DFIGs) is designed considering DRU characteristics. By coordinating the power outputs of DFIGs and hydrogen production units, the system enables both wind energy utilization and inertia support. Simulation results show that, by applying the proposed method, the offshore wind farm connected to DRU can operate reliably and provide the required active power support to the onshore grid. The proposed strategy effectively enhances reliability and flexibility of DRU-based wind farms with hydrogen production, offering a promising solution for deep-sea renewable integration.**

Keywords-offshore wind farm; hydrogen production; diode rectifier

I. INTRODUCTION

As the world advances toward renewable energy adoption and carbon neutrality, offshore wind power has experienced rapid growth due to its abundant resources and stable output. It is projected that by 2050, the installed capacity of offshore wind power will reach 2000 GW [1]. At the same time, the growing demand for hydrogen as a clean fuel for maritime

applications [2] presents new opportunities for the integrated utilization of offshore wind energy.

Therefore, integrating hydrogen production units (HPU) within offshore wind farms to produce hydrogen on-site enables the local utilization of offshore wind energy. This approach not only facilitates the consumption of offshore wind power but also provides a reliable hydrogen source for hydrogen-powered vessels [3].

To address the inherent intermittency and uncertainty of wind power, Melo and Chang-Chien proposed a coordinated control strategy between offshore wind farms and an onshore hydrogen management system (HMS) [4]. In this scheme, the HMS is responsible for medium- and long-term energy balancing, while a supercapacitor is introduced to provide short-term dynamic compensation. Besides, as commercial electrolyzers possess limited power ramping capabilities, mismatches between fluctuating wind output and electrolyzer demand may lead to frequent disconnections and operational instability. Timmers, Egea-Àlvarez, Gkountaras and Xu proposed an off-grid operation scheme that integrates energy storage, rotor inertia, and pitch control [5]. The study also points out that improving the response speed of the electrolyzer can significantly reduce—or even eliminate—the need for energy storage. In order to address energy balancing and grid congestion issues arising from large-scale offshore wind integration, Mu, Yang, Sun and Larumbe proposed the wind farm co-locating with a hydrogen electrolysis plant and controlling the electrolyzer to track wind generation in real

979-8-3315-2490-6/25 $31.00 © 2025 IEEE

time [6], which reduces wind power intermittency and limits grid capacity needs.

As nearshore wind resources become increasingly saturated, offshore wind farms are expanding into deeper-sea regions. The low-cost and highly reliable offshore converter stations have become a research focus. Among various solutions, the diode rectifier unit (DRU) based HVDC scheme has emerged as a promising option and has attracted growing attention [7].

Due to the uncontrollable nature of the DRU, offshore wind turbines connected to DRU must operate in a grid-forming mode. To achieve synchronized operation among wind turbines, communication links [8] and GPS [9] are used to transmit phase and frequency information. With the purpose of avoiding the need for additional hardware, a synchronization control strategy based on a phase-locked loop (PLL) has been proposed [10]. Furthermore, leveraging the self-synchronization mechanism of the DRU, P-V/Q-f control loops have been designed to eliminate the stability issues introduced by the PLL [11].

However, the existing researches on DRU-based systems focus primarily on PMSGs and does not consider the impact of wind turbines and hydrogen production units integration on wind farm control strategies. Hence, for DFIG-based offshore wind farm with hydrogen production connected to DRU-HVDC, whose configuration is illustrated in Fig. 1, this paper proposes a grid-forming control strategy for the DFIG connected to the DRU-HVDC. Based on this strategy, coordinated power control between the DFIGs and hydrogen production units is implemented to maximize wind energy utilization and enhance the support capability of the DRU-HVDC system to the onshore grid.

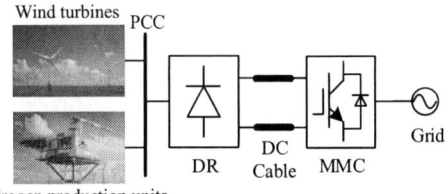

Fig. 1 The configuration of the DFIG-based offshore wind farm with hydrogen production connected to DRU-HVDC.

II. CONTROL OF DFIG AND HYDROGEN PRODUCTION UNIT

A. Grid-Forming Control of the DFIG

Due to the uncontrollability of DRU, DFIG needs to operate in the grid-forming mode. In addition, the self-synchronization mechanism of the DRU-based wind farm, which can be achieved by P-V and Q-f control loops, is different from that of the conventional grid, so the grid-forming control strategy of DFIG needs to be redesigned.

Since the power of DFIG is mainly output from the stator side, the rotor-side converter (RSC) adopts the grid-forming control strategy, as shown in Fig. 2. The grid-side converter (GSC) is ignored.

1) Power control loop

To achieve self-synchronization and reactive power sharing of different DFIGs, the Q-f control loop needs to be designed. The droop controller can meet the requirement of power sharing, so the Q-f control loop can be given by

$$\omega_1^* = \omega_0 - \frac{k_g}{1+sk_t}(Q_s^* - Q_s) \tag{1}$$

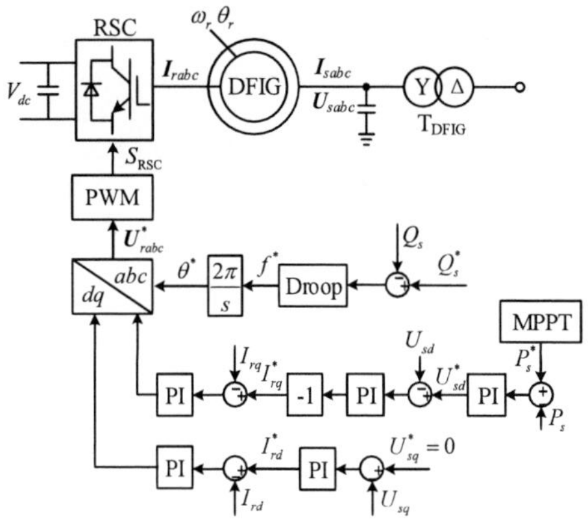

Fig. 2 The proposed grid-forming control of the DFIG connected to DRU-HVDC.

where ω_1^* and Q_s^* denote the angular frequency reference of the voltage and the reference of the reactive power output by the DFIG's stator; k_g and k_t represent the proportional and time parameters of the droop controller, respectively; Q_s represents the reactive power output by the DFIG's stator; ω_0 represents the initial value of the angular frequency.

According to (1), the angle used to coordinate transformation can be calculated as

$$\theta_1 = \omega_1^* \omega_{base}/s \tag{2}$$

where ω_{base} represents the basic value of the angular frequency.

Then, the P-V control loop can be realized by a PI controller, which can be expressed as

$$\begin{cases} U_{sd}^* = (k_{Pp} + k_{Pi}/s)(P_s^* - P_s) \\ U_{sq}^* = 0 \end{cases} \tag{3}$$

where U_{sd}^* and U_{sq}^* denote the amplitude reference of the d- and q-axis stator voltage; P_s^* represents the reference of the active power output by the DFIG's stator; k_{Pp} and k_{Pi} represent the proportional and integral parameters of the PI controller, respectively; P_s represents the active power output by the DFIG's stator.

2) Voltage control loop

The three-phase stator voltages can be converted into the corresponding dq-axis stator voltages using (2). The PI controller is selected to track the stator voltages and generate the rotor current references.

$$\begin{cases} I_{rd}^* = (k_{Up} + k_{Ui}/s)(U_{sq}^* - U_{sq}) \\ I_{rq}^* = -(k_{Up} + k_{Ui}/s)(U_{sd}^* - U_{sd}) \end{cases} \tag{4}$$

979-8-3315-2490-6/25 $31.00 © 2025 IEEE

where I_{rd}^* and I_{rq}^* denote the amplitude reference of the d- and q-axis rotor current; k_{Up} and k_{Ui} represent the proportional and integral parameters of the PI controller, respectively; U_{sd} and U_{sq} represents the d- and q-axis stator voltages.

3) Current control loop

The three-phase rotor currents can be converted into the corresponding dq-axis rotor currents using (2). The PI controller is selected to track the rotor currents and generate the rotor voltage references.

$$\begin{cases} U_{rd}^* = (k_{Ip} + k_{Ii} / s)(I_{rd}^* - I_{rd}) \\ U_{rq}^* = (k_{Ip} + k_{Ii} / s)(I_{rq}^* - I_{rq}) \end{cases} \quad (5)$$

where U_{rd}^* and U_{rq}^* denote the amplitude reference of the d- and q-axis rotor voltage; k_{Ip} and k_{Ii} represent the proportional and integral parameters of the PI controller, respectively; I_{rd} and I_{rq} represents the d- and q-axis rotor currents.

Based on (5), the switching signals for the rotor-side converter can be obtained by using PWM technology.

B. Control of the Hydrogen Production Unit

Since the hydrogen production units need to participate in the power regulation of offshore wind farm, PWM converter, which is capable of safe operation over a wide range, is selected as the rectification circuits for the hydrogen production device. And the control structure is exhibited in Fig. 3.

To obtain phase information of offshore AC grid, phase-locked loop (PLL) is employed, which can be described as

$$H_{Hpll}(s) = (k_{Hpllp} + k_{Hplli} / s) / s \quad (6)$$

where k_{HPLLp} and k_{HPLLi} represent the proportional and integral parameters of the PI controller, respectively; U_{sd} and U_{sq} represents the d- and q-axis stator voltages.

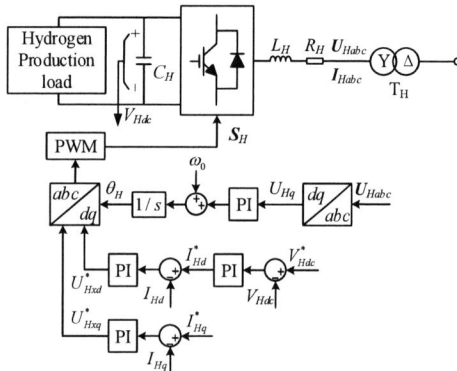

Fig. 3 The control of the hydrogen production unit.

The PI controller is selected to track the DC-link voltage and generate the current references.

$$\begin{cases} I_{Hd}^* = (k_{HVp} + k_{HVi} / s)(V_{Hdc}^* - V_{Hdc}) \\ I_{Hq}^* = 0 \end{cases} \quad (7)$$

where I_{Hd}^* and I_{Hq}^* denote the reference of the PWM converter's d- and q-axis current; k_{HVp} and k_{HVi} represent the proportional and integral parameters of the PI controller,

respectively; V_{Hdc}^* and V_{Hdc} represents the reference and the actual value of DC-link voltage.

The current control loop of the PWM converter can be realized by a PI controller, which can be expressed as

$$\begin{cases} U_{Hxd}^* = (k_{HIp} + k_{HIi} / s)(I_{Hd}^* - I_{Hd}) \\ U_{Hxq}^* = (k_{HIp} + k_{HIi} / s)(I_{Hq}^* - I_{Hq}) \end{cases} \quad (8)$$

where U_{Hxd}^* and U_{Hxq}^* denote the amplitude reference of the PWM converter's d- and q-axis voltage; k_{HIp} and k_{HIi} represent the proportional and integral parameters of the PI controller, respectively; I_{Hd} and I_{Hq} represents the d- and q-axis currents of the PWM converter.

Based on (8), the switching signals for the PWM converter can be obtained by using PWM technology.

III. COORDINATED CONTROL OF OFFSHORE WIND FARM

Offshore wind farms equipped with hydrogen production units not only provides a hydrogen source for hydrogen-powered ships but also introduces an additional path of wind power. Therefore, a coordinated power control strategy is required to enable efficient utilization of offshore wind energy.

A. Maintain the Power Transmitted by DRU

When the power demand of the onshore load is fixed, the power transmitted by the DRU should remain constant. However, during this period, wind speed fluctuations may occur, leading to variations in the power outputs of the DFIGs. To maintain a stable power transmitted by the DRU, the power consumption of the hydrogen production units needs to be adjusted accordingly based on the changes of DFIG power outputs.

Assume the offshore wind farm consists of m DFIGs and n hydrogen production units. The active power outputs of the DFIGs are denoted as P_{DFIG1}, P_{DFIG2}, ..., P_{DFIGm}. The active power consumed by the hydrogen production units are denoted as P_{H1}, P_{H2}, ..., P_{Hn}.

The variation in active power outputs of the DFIGs under wind speed fluctuations is given by

$$\Delta P_{DFIG} = \sum_{i=1}^{m} \Delta P_{DFIGi} \quad (9)$$

where ΔP_{DFIGi} represents the active power change of the i-th DFIG due to wind speed variation.

The variation in active power outputs of the DFIGs is allocated to each hydrogen production unit in proportion to its available dispatchable capacity, in order to maintain the power transmitted by the DRU. The variation in active power consumed by each hydrogen production unit can be derived as

$$\Delta P_{H1} : \Delta P_{H2} : ... : \Delta P_{Hn} = P_{Hd1} : P_{Hd2} : ... : P_{Hdn} \quad (10)$$

where ΔP_{Hi} represents the active power change of the i-th hydrogen production unit; P_{Hdi} represents the dispatchable power of the i-th hydrogen production unit.

Taking the case of an increase in active power outputs of the DFIGs as an example, in order to ensure that the power transmitted by the DRU remains constant, the hydrogen production units need to absorb this additional power as much as possible. In proportion to the dispatchable power of the hydrogen production units, which is defined as the difference between its rated power and its actual operating power, the power reference of each hydrogen production unit is increased, thereby converting the excess wind energy into hydrogen and minimizing wind energy curtailment.

B. Frequency Support for the Onshore Grid

Domestic and foreign grid codes require wind farms to have the capability to participate in frequency regulation of the power system. When frequency deviations occur in the onshore grid, offshore wind farms must adjust the power output to help suppress the frequency fluctuations.

Wind farms can participate in frequency regulation by incorporating supplementary control strategies in response to frequency deviations, including inertia control and droop control. The inertia control can be written as

$$\Delta P_{in} = -k_{in} \frac{df}{dt} \tag{11}$$

where ΔP_{in} represents the inertial support power provided by the wind farm; k_{in} represents the control parameter of the inertia control; f represents the frequency of the onshore grid.

The droop control can be described as

$$\Delta P_{dr} = k_{dr}(f_0 - f) \tag{12}$$

where ΔP_{dr} represents the droop support power provided by the wind farm; k_{dr} represents the control parameter of the droop control; f_0 represents the rated frequency of the onshore grid.

Therefore, the total power contributed by the offshore wind farm for onshore grid frequency support can be given by

$$\Delta P_{su} = \Delta P_{in} + \Delta P_{dr} \tag{13}$$

Considering the power regulation characteristics of hydrogen production units, they are not suitable for participating in inertia control to provide inertial support power. Therefore, this portion of the power must be fully supplied by the DFIGs. In contrast, for droop control, the required power adjustment can be shared between the DFIGs and the hydrogen production units.

IV. SIMULATION RESULTS

To verify the effectiveness of the proposed control strategy, a case study of an offshore wind farm equipped with hydrogen production units and transmitted via a DRU was developed in the MATLAB/Simulink. As shown in Fig. 4, the system includes three aggregated DFIGs rated at 330 MW each, two aggregated hydrogen production units rated at 100 MW each, a DRU-based offshore converter station, and an MMC-based onshore converter station. Besides, the parameters of the system are listed in Table I.

A. Grid-Forming Control of DFIGs

The simulation results of the grid-forming control are presented in Fig. 5. Active power references are assigned to the

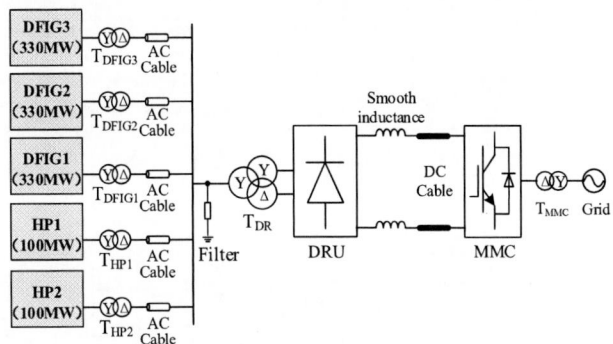

Fig. 4 The diagram of the offshore wind farm system.

TABLE I. THE PARAMETER OF THE SYSTEM

	Items	Value
DFIG	Rated capacity	330MW
	Rated AC voltage	690V
	Pole pairs	2
	Stator and rotor resistance	R_s=0.026p.u., R_r=0.022p.u.
	Stator and rotor leakage inductance	$L_{s\sigma}$=0.173p.u., $L_{r\sigma}$=0.167p.u.
	Mutual inductance	2.78p.u.
Hydrogen Production Unit	Rated capacity	100MW
	Rated AC voltage	690V
	Filter resistance and inductance	L_{lf}=1.568p.u., R_{lf}=0.002p.u.
	Rated DC voltage	1200V
Offshore AC Grid	AC cable	R_{ac}=0.6Ω, L_{ac}=2mH, C_{ac}=2×0.672μF
	Turns ratio of T_{DFIG} and T_{HP}	0.69kV/66kV
	Inductance of T_{DFIG} and T_{HP}	0.08p.u.
DRU	Rated AC voltage	66kV
	Rated DC voltage	640kV
	Turns ratio of T_{DR}	66kV/261.8kV/261.8kV
	Inductance of T_{DR}	0.18p.u.
	Filter of DRU	C_1=116.98μF; C_2=116.98μF, C_3=1300μF, L_1=7.8mH, R_1=1.7Ω, R_2=14.97Ω; C_4=116.98μF, L_2=0.78mH, R_3=4.76Ω
	DC smoothing reactance	240mH
DC Line	DC cable	R_{dc}=1.25Ω, L_{dc}=60mH, C_{dc}=26μF
MMC	Rated capacity	1000MW
	Rated DC voltage	640kV
	Number of SMs per arm	200
	SM capacitance	6510μF
	Rated capacitor voltage	3.2kV
	Arm induction	77.8mH
	Turns ratio of T_{MMC}	230kV/330kV
Onshore AC Grid	Rated AC voltage	230kV

DFIG1, DFIG2, and DFIG3 at 0.5 s, 1.5 s, and 2.5 s, respectively. It can be observed that, adopting the proposed grid-forming control strategy, each DFIG is able to start

979-8-3315-2490-6/25 $31.00 © 2025 IEEE

smoothly and accurately track its active power reference. Moreover, the three DFIGs achieve self-synchronization and maintain stable operation of the offshore wind farm.

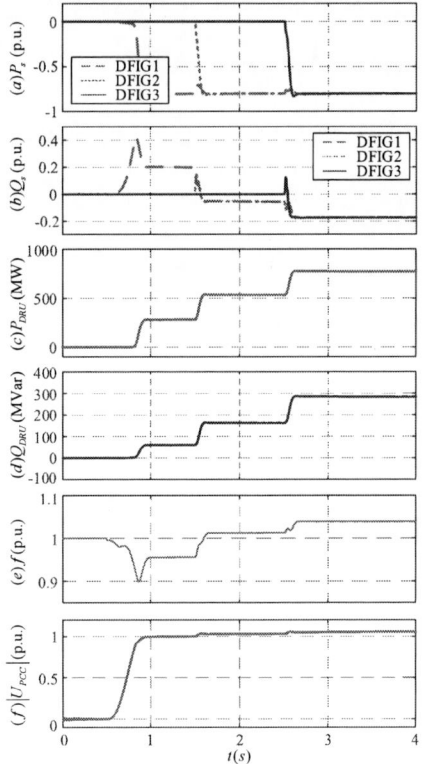

Fig. 5 The simulation results of grid-forming control: (a) Active power of DFIGs; (b) Reactive power of DFIGs; (c) Active power of DRU; (d) Reactive power of DRU; (e) Frequency of offshore wind farm; (f) Voltage of offshore wind farm.

B. Power Maintenance

The simulation results of power maintenance are shown in Fig. 6. At 4.5 s, wind speeds change for DFIG1 and DFIG2, causing the active power outputs to increase to 1 p.u. and 0.9 p.u., respectively. According to the proposed control strategy, the two hydrogen production units adjust each power consumption proportionally based on each dispatchable capacities. As shown in Fig. 6(e), after a brief adjustment process, the active power transmitted by the DRU remains constant, effectively avoiding wind energy curtailment.

To further verify the robustness of the proposed method, simulation results of power maintenance with wind speed decrease are shown in Fig. 7. As the wind speeds of DFIG1 and DFIG2 decrease, the active power outputs of DFIG1 and DFIG2 drop to 0.7 p.u. To maintain constant active power transmission through the DRU, the power consumption of the hydrogen production units is correspondingly reduced. As shown in Fig. 7(e), the proposed control strategy effectively maintains the active power delivered to the onshore grid.

C. Frequency Support

The simulation results of frequency support are exhibited in Fig. 8. At 4.5 s, the onshore grid frequency drops abruptly from

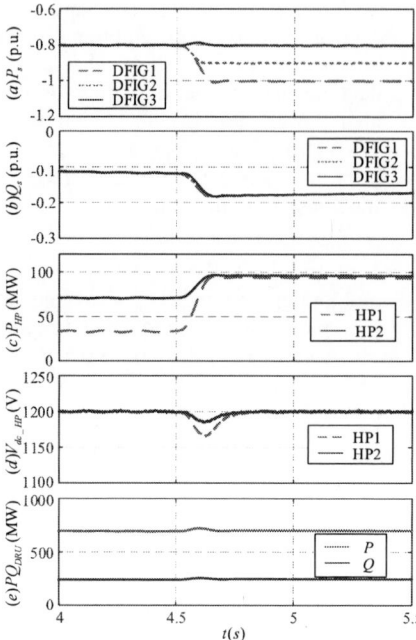

Fig. 6 The simulation results of power maintenance with wind speed increase: (a) Active power of DFIGs; (b) Reactive power of DFIGs; (c) Active power of hydrogen production units; (d) DC-link voltage of hydrogen production units; (e) Active and reactive power of DRU.

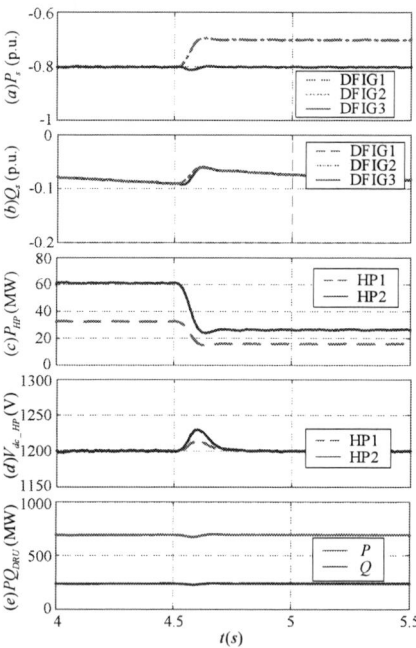

Fig. 7 The simulation results of power maintenance with wind speed decrease: (a) Active power of DFIGs; (b) Reactive power of DFIGs; (c) Active power of hydrogen production units; (d) DC-link voltage of hydrogen production units; (e) Active and reactive power of DRU.

50 Hz to 49.5 Hz. Since only the DFIGs employ inertia control, the three DFIGs respond rapidly in the initial stage of the frequency disturbance, providing immediate frequency support. As the rate of frequency change decreases, the additional power

contribution from inertia control diminishes. When the deviation between the actual and rated frequency becomes larger, the droop control becomes dominant. The power consumption of the two hydrogen production units gradually decreases, while the additional active power output from the three DFIGs stabilizes under the influence of droop control. These results demonstrate that the proposed control strategy enables coordinated power regulation between the DFIGs and hydrogen production units, effectively providing enhanced power support during onshore frequency deviations.

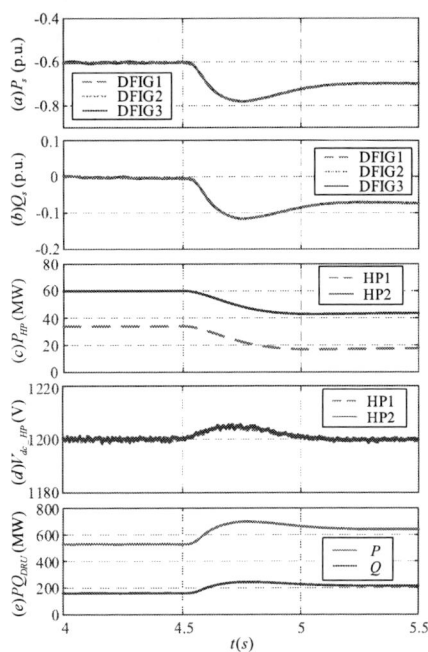

Fig. 8 The simulation results of frequency support: (a) Active power of DFIGs; (b) Reactive power of DFIGs; (c) Active power of hydrogen production units; (d) DC-link voltage of hydrogen production units; (e) Active and reactive power of DRU.

V. DISCUSSION

This study aims to develop a robust control scheme for DFIG-based offshore wind farms with hydrogen units connected to DRU-based HVDC systems. Due to the uncontrollability of DRU, wind turbines connected to DRU need to operate in grid-forming mode to construct the offshore AC power grid. According to the simulation results in Fig. 5, it can be seen that the proposed grid-forming control strategy can ensure the safe and reliable operation of DFIG connected to DRU, which also provides AC voltage for the normal operation of hydrogen production units.

On this basis, considering the uncertain factors during operation, including wind speed changes, onshore load changes, etc., the power of DFIGs and hydrogen production units in offshore wind farms should be reasonably regulated to achieve different control objectives.

Compared to existing researches, this paper, for the first time, proposes the application of DFIG and hydrogen production equipment in the DRU-based HVDC system, and

proposes corresponding collaborative control strategies to better support the onshore power grid.

In addition, the control for DRU under faults is critical. For PMSG-based wind farms connected to DRU, Li proposed voltage and current limiting control strategies for various fault types, combined with relay protection schemes to ensure reliable operation under fault conditions [12]. Based on this, for the wind farm configuration discussed in this paper, if the wind turbines are still able to maintain the offshore AC grid, the hydrogen production units remain connected and adjust the power references according to the wind farm's operational requirements. However, if the fault is severe enough to cause disconnection of the wind turbines, the hydrogen production units must also be disconnected.

VI. CONCLUSION

This paper proposes a DRU-based offshore wind farm with hydrogen production aimed at reducing the development cost of deep-sea wind power and addressing the hydrogen supply challenge for hydrogen-powered ships. To support this scheme, a grid-forming control strategy for DFIGs connected to DRU-based transmission is developed. On this basis, coordinated power control of DFIGs and hydrogen production units is implemented to achieve various control objectives. Simulation results demonstrate that the proposed control strategy can reliably enable both wind power transmission and hydrogen production, maximizing wind energy utilization and enhancing the capability of the offshore wind farm to support onshore grid. The stability analysis of the DRU-based offshore wind farm with hydrogen production will be investigated in near future.

ACKNOWLEDGMENT

The authors of this paper acknowledge the financial support from the National Key R&D Program of China (2023YFB4204700), CHNG science and technology project (HNKJ22-H146).

REFERENCES

[1] GWEC, "Global Offshore Wind Report 2024," 2024. [Online]. Available: https://gwec.net/global-offshore-wind-report-2024/.

[2] M. Jamma, P. Bujlo and Ø. Ulleberg, "Intelligent energy management system for hybrid fuel cell/battery powered ships," IEEE Trans. Ind. Appl., doi: 10.1109/TIA.2025.3571358. (Early Access)

[3] D. Jang, K. Kim, K. Kim, and S. Kang, "Techno-economic analysis and Monte Carlo simulation for green hydrogen production using offshore wind power plant," Energy Convers. Manage., vol. 263, pp. 115695, July 2022.

[4] D. F. Recalde Melo and L. -R. Chang-Chien, "Synergistic control between hydrogen storage system and offshore wind farm for grid operation," IEEE Trans. Sustain. Energy, vol. 5, no. 1, pp. 18-27, January 2014.

[5] V. Timmers, A. Egea-Àlvarez, A. Gkountaras and L. Xu, "Control and power balancing of an off-grid wind turbine with co-located electrolyzer," IEEE Trans. Sustain. Energy, vol. 15, no. 4, pp. 2349-2360, October 2024.

[6] H. Mu, D. Yang, Y. Sun and L. B. Larumbe, "Dynamic power tracking performance and small signal stability analysis of integrated wind-to-hydrogen system," IEEE Trans. Sustain. Energy, vol. 15, no. 4, pp. 2444-2456, October 2024.

[7] C. Wang, A. Ali and F. Blaabjerg, "Composition and control of a new type of hybrid voltage-source converter based on DRUs and FB-MMC

for large-scale offshore wind power integration and transmission," IEEE Trans. Power Electron., vol. 39, no. 5, pp. 5721-5732, May 2024.

[8] R. Blasco-Gimenez, S. Añó-Villalba, J. Rodríguez-D'Derlée, F. Morant, and S. Bernal-Perez "Distributed voltage and frequency control of offshore wind farms connected with a diode-based HVdc link," IEEE Trans. Power Electron., vol. 25, no. 12, pp. 3095–3105, December 2010.

[9] C. Prignitz, H.-G. Eckel, and A. Rafoth, "FixReF sinusoidal control in line side converters for offshore wind power generation," in2015 IEEE 6th International Symposium on Power Electronics for Distributed Generation Systems (PEDG), Aachen, Germany: IEEE, June 2015, pp. 1–5.

[10] L. Yu, R. Li, and L. Xu, "Distributed PLL-based control of offshore wind turbines connected with diode-rectifier-based HVDC systems," IEEE Trans. Power Del., vol. 33, no. 3, pp. 1328–1336, June 2018.

[11] H. Xiao, X. Huang, Y. Huang and Y. liu, "Self-synchronizing control and frequency response of offshore wind farms connected to diode rectifier based HVDC system," IEEE Trans. Sustain. Energy, vol. 13, no. 3, pp. 1681–1692, July 2022.

[12] R. Li, "Operation of offshore wind farms connected with DRU-HVDC transmission systems with special consideration of faults," Glob. Energy Interconnect., vol. 1, no. 5, 2018.

2025 International Conference on Advanced Energy Systems and Power Electronics (AESPE 2025)

Thermal-Hydraulic Analysis of 7-Rod Bundle Fuel Assembly in Lead-Cooled Fast Reactor

Hui Tianyu*
Suzhou Nuclear Power Research Institute,
Suzhou, Jiangsu Province, China
Corresponding author e-mail: 2417245892@qq.com

Ying Hong
Suzhou Nuclear Power Research Institute,
Suzhou, Jiangsu Province, China
e-mail: yinghong94@126.com

Liu Tao
Suzhou Nuclear Power Research Institute,
Suzhou, Jiangsu Province, China
e-mail: liut@cgnpc.com.cn

Wang Qiang
Suzhou Nuclear Power Research Institute,
Suzhou, Jiangsu Province, China
e-mail: 3947646786@qq.com

Fang Kewei
Suzhou Nuclear Power Research Institute,
Suzhou, Jiangsu Province, China
e-mail: 673520207@qq.com

Wang Binfeng
Suzhou Nuclear Power Research Institute,
Suzhou, Jiangsu Province, China
e-mail: wangbinfeng88@126.com

Tang Tang
Suzhou Nuclear Power Research Institute,
Suzhou, Jiangsu Province, China
e-mail: 1784115161@qq.com

Lin Nan
Suzhou Nuclear Power Research Institute,
Suzhou, Jiangsu Province, China
e-mail: linnan216@163.com

Abstract—**Thermal-hydraulic analysis of fuel assemblies in lead-cooled fast reactors plays a crucial role in ensuring safe and efficient reactor operation. To precisely characterize the three-dimensional flow field, temperature distribution, and pressure drop characteristics within the fuel assemblies, high-fidelity three-dimensional numerical simulations were conducted. This study employs computational fluid dynamics (CFD) methodology, focusing on a hexagonal 7-rod fuel assembly configuration. A comparative analysis was performed between wire-wrapped and bare-rod assemblies to investigate the effects of helical wire spacers on coolant flow patterns, heat transfer enhancement, and pressure drop characteristics. The simulation results demonstrate excellent agreement between the CFD-predicted temperature profiles, velocity distributions, pressure drop characteristics and theoretical analyses, thereby validating the computational accuracy and reliability of the established numerical model. These findings preliminarily demonstrate that the adopted simulation approach can provide valuable references for thermal-hydraulic optimization of lead-cooled fast reactor fuel assemblies. The results demonstrate that wire-wrapped assemblies provide better coolant mixing and heat transfer than bare rods, despite slightly higher pressure drops. This study offers valuable guidance for optimizing lead-cooled fast reactor fuel assembly designs. The validated CFD methodology contributes essential data for advancing next-generation reactor development.**

Keywords- Lead-Cooled Fast Reactor; CFD; fuel assemblies;

I. INTRODUCTION

Under the impetus of global energy transition and carbon neutrality goals, nuclear energy, as a stable and reliable clean baseload energy source, plays an irreplaceable strategic role in optimizing energy structures. Compared to intermittent renewable energy sources such as wind and solar power, nuclear energy has distinct advantages, including high energy density and stable output, making it currently the most promising low-carbon energy option for large-scale fossil fuel replacement[1]. China's "thermal reactor-fast reactor-fusion reactor" three-step development strategy positions fast reactor technology as a key component for the sustainable development of nuclear energy, with its core objectives being nuclear fuel breeding and high-level waste transmutation—fundamentally improving nuclear fuel utilization efficiency and addressing nuclear waste disposal challenges[2].

Among various fast reactor technologies, lead-cooled fast reactors (LFRs) are regarded as one of the most promising options due to their superior safety and economic viability. Utilizing lead/lead-bismuth alloy as a coolant, this reactor type inherently avoids the chemical reactivity risks associated with sodium-cooled systems while offering high thermal conductivity and a high boiling point, significantly enhancing inherent reactor safety. Notably, its high thermal inertia and strong natural circulation capability enable effective heat removal even under accident conditions, providing a new safety paradigm for next-generation reactor designs[3].

979-8-3315-2490-6/25 $31.00 © 2025 IEEE

The thermal-hydraulic characteristics of fuel assemblies are critical to the design and safety of lead-cooled fast reactors. Due to the high density and low viscosity of lead-based coolants, the flow and heat transfer processes within fuel assemblies differ markedly from those in conventional water-cooled or sodium-cooled reactors. Furthermore, the wire-wrapped spacer design alters coolant flow distribution, affecting local heat transfer and flow resistance. Insufficient heat transfer or flow instabilities may lead to localized overheating or structural failure, jeopardizing reactor safety[4][5]. Therefore, in-depth investigation of the complex thermal-hydraulic behavior in LFR fuel assemblies is essential for guiding engineering design. This study employs computational fluid dynamics (CFD) methods to analyze the influence of wire-wrapped spacers on flow and heat transfer characteristics within hexagonal fuel assemblies of lead-cooled fast reactors.

II. COMPUTATIONAL MODEL AND BOUNDARY CONDITIONS

The study selects a seven-pin hexagonal fuel assembly as the research object, with its geometric configuration illustrated in Figure 1.

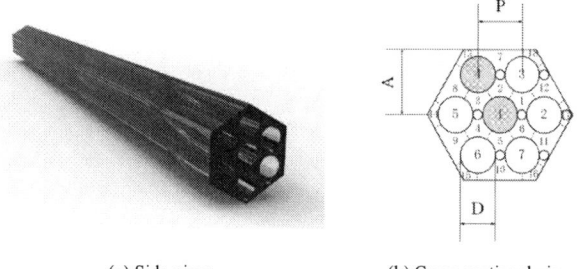

(a) Side view   (b) Cross-sectional view

Figure 1. Schematic Diagram of Seven-Rod Fuel Assembly Structure

In this computational model, LBE is employed as the coolant. The modeling approach focuses solely on the fluid region, excluding the solid domains (such as cladding and fuel pellets) to simplify the analysis while maintaining accuracy. The geometric dimensions and boundary conditions of the seven-rod fuel assembly are presented in TABLE I[6][7]. The boundary conditions applied in the simulation, along with the thermophysical properties of the LBE coolant, are detailed in TABLE II[6][7]. These include parameters such as inlet velocity, temperature profiles, and pressure conditions, ensuring a comprehensive representation of the physical behavior in the system.

TABLE I.   PARAMETERS OF PARALLEL THREE-CHANNEL PROBLEM SYSTEM

Parameters	Value	Unit
Geometric Dimensions of the Computational Model		
Pitch	10.49	mm
Diameter	8.2	mm
Height	450	mm
Assembly Duct Flat-to-Flat Distance	31.0	mm
Wire Wrap Diameter	2.2	mm
Axial Pitch	150	mm
Boundary Conditions		
Mass Flow Rate	7.5/10/12.5	kg/s
Inlet Temperature	473.15	K
Total Thermal Power	200	kW
Assembly Duct/Heater Rod Surface	No-Slip Condition	

TABLE II.   PROPERTIES OF LEAD-BISMUTH ALLOY

Parameters	Value	Unit
Density	10469.74	kg/m^3
Specific Heat Capacity	148.63	J/kg/K
Dynamic Viscosity	0.00253	Pa · s
Thermal Conductivity	9.75	W/m/K

III. MESH GENERATION AND SENSITIVITY ANALYSIS FOR NON-WIRE-WRAPPED ASSEMBLIES

Based on the ANSYS Mesh, six mesh generation schemes were designed for the hexagonal 7-rod fuel assembly, as listed in TABLE III. The fluid domain geometry and mesh models were established in Fluent, as shown in Figure 2.

TABLE III.   MESH DESIGN SCHEME

Mesh Scheme	Radial Mesh	Axial Layers	Total Grid
1	930	20	18600
2	930	25	23250
3	930	30	27900
4	1625	20	32500
5	1625	25	40625
6	1625	30	48750

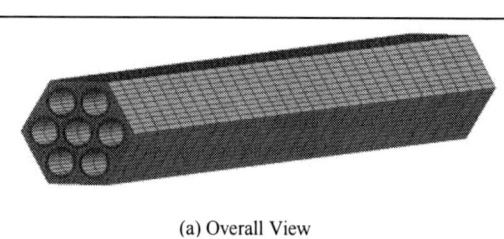

(a) Overall View

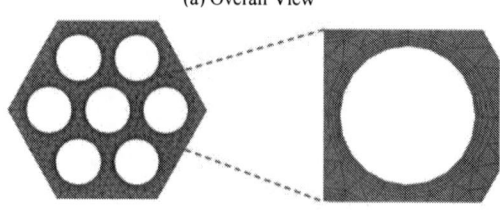

(b) Enlarged Detail View

Figure 2. Schematic Diagram of Model Mesh Generation

In the simulation, the standard high-Reynolds-number k-ε turbulence model was employed. This model is based on the transport equations for turbulent kinetic energy (k) and its dissipation rate (ε). Although primarily designed for high-Reynolds-number flows, it also accounts for low-Reynolds-number effects to a certain extent.

To meet the requirements of the selected turbulence model (ensuring y+ ≈ 30), four boundary layers were adopted, as illustrated in Figure 2.

The unwired fuel assembly was selected as the research object. Grid independence verification was conducted using the outlet average temperature, outlet average pressure drop, and the axial surface temperature of the central channel as evaluation metrics. The results are presented in Figures 3 and 4. As shown in Figures 3 and 4, the six mesh schemes yield similar results in temperature distribution while showing minor differences in pressure drop calculation. To ensure computational accuracy, this study adopts the sixth mesh scheme, which possesses a higher proportion of orthogonal grids.

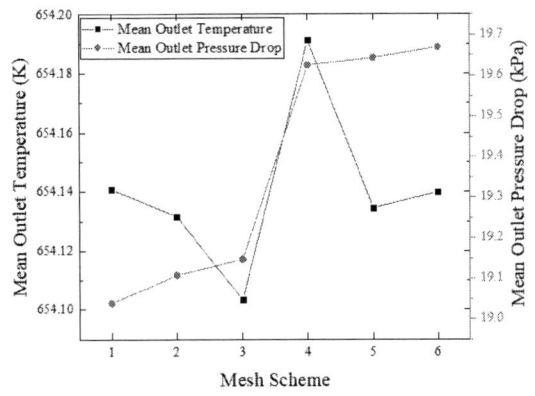

Figure 3. Variations in Mean Outlet Temperature and Pressure Drop Across Different Meshing Schemes

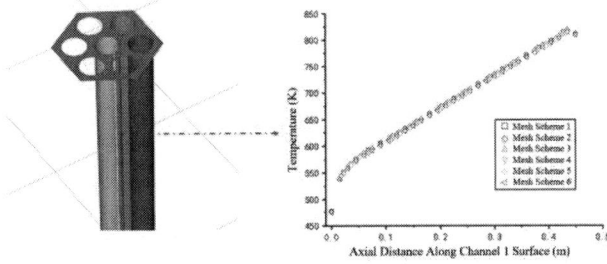

(a) Extraction Path Location on Channel 1 Surface (b) Temperature Distribution Along Sampling Lines Under Different Mesh Schemes

Figure 4. Temperature Distribution on the Central Channel Surface for Different Mesh Schemes

IV. RESULTS AND DISCUSSION

This section focuses on the numerical simulation analysis of non-wire-wound/wire-wound assemblies based ANSYS Fluent with the SIMPLE (Semi-Implicit Method for Pressure-Linked Equations) solver, including: inlet and outlet pressure drop and

average temperature, temperature/velocity distributions at different axial heights and surface temperature distribution along the channel.

A. Flow and Heat Transfer Characteristics of Non-Wire-Wound Assemblies

1) Inlet/Outlet Average Pressure Drop and Temperature

According to the boundary conditions of the computational model given in TABLE I, the calculation results for the non-wire-wound 7-rod fuel assembly after 2000 iterations are presented in TABLE IV. and Figure 5.

TABLE IV. COMPARISON OF NUMERICAL RESULTS WITH THEORETICAL VALUES

Inlet Flow Rate	Outlet Average Temperature			Outlet Average Pressure Drop
	Caculated Value	Theoretical Value	Error	Caculated Value
7.5 kg/s	654.18 K	652.57 K	1.61 K	19761 Pa
10 kg/s	609.03 K	607.71 K	1.32 K	32891 Pa
12.5 kg/s	581.92 K	580.80 K	1.12 K	48832 Pa

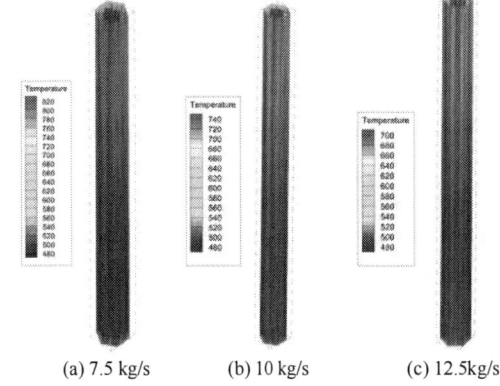

(a) 7.5 kg/s   (b) 10 kg/s   (c) 12.5kg/s

Figure 5. Thermal Contour Analysis of Non-Wire-Wrapped Fuel Assemblies (Left:; Middle: 10 kg/s; Right: 12.5 kg/s)

The computational model in this report shows good agreement with theoretical values, with a maximum temperature deviation not exceeding 2 K. The temperature distribution of the non-wire-wrapped fuel assembly exhibits higher values in the central channels and lower values on both sides, which aligns with fundamental physical principles. Therefore, it can be concluded that the results obtained from the non-wire-wrapped model developed in this report are reasonably reliable.

2) Cross-Sectional Velocity/Temperature Contour Maps at Different Axial Heights

Taking the 12.5 kg/s condition as the study case, cross-sectional velocity/temperature contour maps at different axial heights are obtained, with the results shown in Figure 6.

As shown in Figure 6, as the height increases, the temperature rise in the subchannels of the fuel assembly is the fastest, followed by the edge channels, while the corner channels exhibit the slowest temperature increase. This occurs because subchannels are located closer to the heated surfaces of the rod bundles, with a smaller wetted perimeter and a larger heated perimeter, leading to more rapid temperature rise. In

979-8-3315-2490-6/25 $31.00 © 2025 IEEE   81

contrast, edge channels exhibit a higher temperature rise than corner channels due to their longer heated perimeter. Additionally, Figure 6 reveals significant circumferential non-uniformity in all channels, which is attributed to the cold-wall effect of the assembly box wall. Notably, the temperature near the box wall side decreases significantly.

As also observed in Figure 6, along the y-axis (axial direction of the fuel assembly), the flow velocity in edge channels and subchannels is higher than in corner channels. Furthermore, after the fluid flows a certain axial distance, only the axial velocity remains, while the radial velocity becomes negligible. This phenomenon is caused by the inlet/outlet effect of lead-bismuth (LBE) coolant. In the entrance region, the flow remains undeveloped, creating a boundary layer near the walls with a steep velocity gradient, thus inducing a radial velocity component. However, beyond a certain axial distance, the inlet/outlet effect diminishes, the flow fully develops into a fully-developed flow, and the radial velocity decreases nearly to zero.

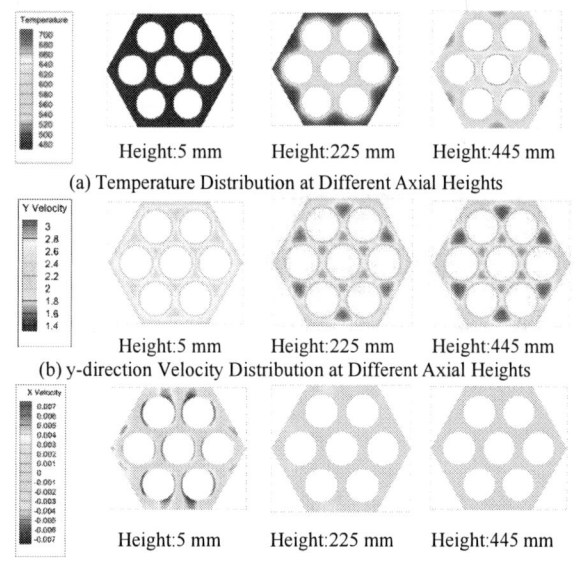

(a) Temperature Distribution at Different Axial Heights

(b) y-direction Velocity Distribution at Different Axial Heights

(c) x-direction Velocity Distribution at Different Axial Heights

Figure 6. Velocity/Temperature Contours at Different Axial Heights

3) Circumferential Temperature Distribution on Heating Rods 1 and 4 at Different Axial Heights

Figure 7 presents the temperature distribution contours of the wire-free assembly at different axial heights. From the figure, it can be observed that in the entrance region, there is no significant circumferential temperature non-uniformity on the surface of the heating rods. However, beyond the 200-mm height, Channel 4 exhibits clear circumferential temperature non-uniformity, and after the 400-mm height, Channel 1 also demonstrates this phenomenon. This behavior occurs because the side closer to the heated surface absorbs more heat, while the side near the assembly box wall absorbs less, leading to circumferential temperature variations. Additionally, theoretical analysis suggests that a smaller pitch (i.e., a reduced distance between adjacent heated surfaces) exacerbates circumferential non-uniformity. Therefore, increasing the pitch

could be an effective strategy to mitigate this non-uniformity in temperature distribution.

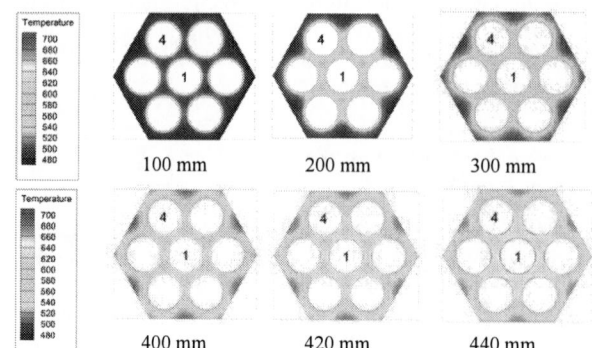

Figure 7. Circumferential Temperature Distribution of Channel 4 and Channel 1

B. Flow and Heat Transfer Characteristics of Wire-Wrapped Assemblies

The geometric model and mesh model of the wire-wrapped assembly are shown in Figure 8. After modeling and meshing, the boundary conditions were kept consistent with those of non-wire-wrapped assemblies, while the k-ε turbulence model and a 4-layer boundary-layer mesh were still adopted.

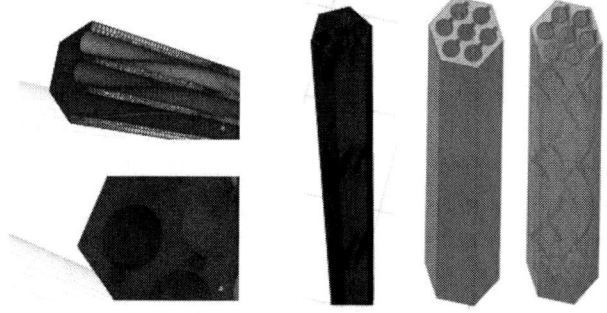

(a) Partial view   (b) Overall View

Figure 8. Schematic of the Wire-Wrapped Assembly Model and Mesh

1) Inlet/Outlet Average Pressure Drop and Temperature

Based on the boundary conditions of the computational model given in TABLE I, the 7-pin wire-wrapped fuel assembly was calculated. After 2000 iterations, the results are presented in TABLE V and Figure 9.

TABLE V. COMPARISON OF NUMERICAL RESULTS WITH THEORETICAL VALUES

Inlet Flow Rate	Outlet Average Temperature			Outlet Average Pressure Drop
	Caculated Value	Theoretical Value	Error	Caculated Value
7.5 kg/s	650.32 K	652.57 K	2.25 K	30408 Pa
10 kg/s	604.62 K	607.71 K	2.09 K	51483 Pa
12.5 kg/s	578.24 K	580.80 K	2.56 K	81334 Pa

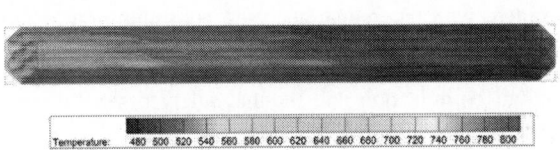

Figure 9. Temperature Distribution Contour (Mass Flow Rate: 12.5 kg/s)

From the results in Table V, it can be seen that the calculated values agree well with the theoretical values, with a maximum deviation not exceeding 3 K. Therefore, the computational results of the unwired model developed in this report can be considered reasonably reliable. Additionally, a comparison with the results for unwired assemblies shows that adding a wire not only increases the additional inlet/outlet pressure drop but also leads to localized high temperatures, approaching 800 K, which is caused by the poor heat transfer between the wire and the heated rod surface.

2) Cross-Sectional Velocity/Temperature Contour Maps at Different Axial Heights

Taking the 12.5 kg/s condition as the study case, cross-sectional velocity/temperature contour maps at different axial heights are obtained, with the results shown in Figure 10.

As shown in Figure 11, a localized high-temperature region appears between the wire spacer and fuel rod due to the low heat transfer coefficient at their interface. The overall temperature distribution still follows a center-peaked pattern, with higher temperatures in the middle and lower temperatures at the sides.

Compared to the unwired assembly, the velocity distribution in the axial direction exhibits significant non-uniformity, and the flow field differs notably from that of the unwired case. Specifically, the axial velocity profile is symmetrical at the inlet section, but as the elevation increases, an asymmetrical distribution develops due to the disruptive effect of the wire spacer on the flow field. In the radial direction, the velocity distribution remains relatively uniform, with minimal influence from the wire spacer. The inlet effects diminish along the flow path, and the overall radial velocity remains small.

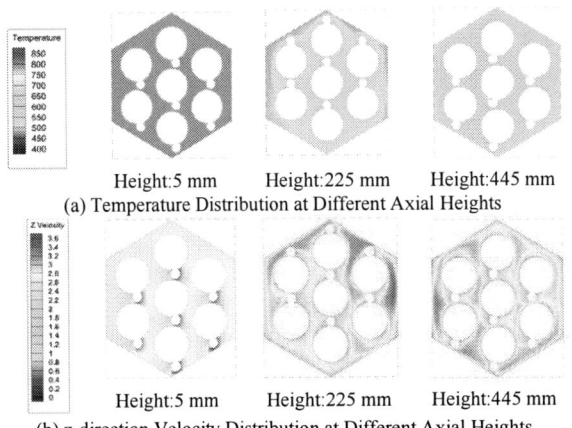

Height:5 mm   Height:225 mm   Height:445 mm
(a) Temperature Distribution at Different Axial Heights

Height:5 mm   Height:225 mm   Height:445 mm
(b) z-direction Velocity Distribution at Different Axial Heights

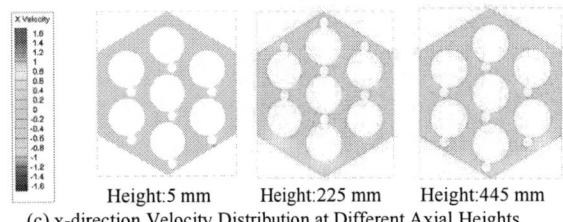

Height:5 mm   Height:225 mm   Height:445 mm
(c) x-direction Velocity Distribution at Different Axial Heights

Figure 10. Velocity/Temperature Contours at Different Axial Heights

3) Circumferential Temperature Distribution on Heating Rods 1 and 4 at Different Axial Heights

Figure 11 presents the temperature contour maps for Channel 1 and Channel 4. As observed, the fuel rod surface temperature exhibits no significant non-uniformity, but the contact interface between the wire spacer and fuel rod shows a much higher temperature compared to the bare rod surface, indicating a pronounced thermal gradient.

To further analyze the circumferential temperature variations, cross-sectional profiles of surface temperatures at different elevations along Channels 1 and 4 were extracted, with results shown in Figure 12. Based on Figure 12, compared to the non-wire-spacer assembly, localized hotspots emerge at different elevations on both the wire spacers and fuel rod surfaces. The circumferential temperature distribution along the wire spacers is notably higher, and the area of these localized high-temperature zones expands with increasing elevation. Overall, the temperature distribution remains stratified: highest near the fuel rod surface and lowest near the assembly box wall.

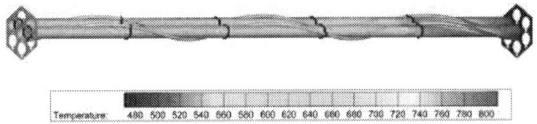

Figure 11. Temperature Distribution in Channel 1 and Channel 4

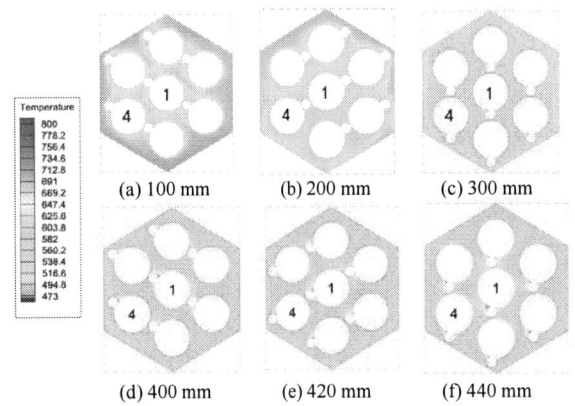

(a) 100 mm   (b) 200 mm   (c) 300 mm

(d) 400 mm   (e) 420 mm   (f) 440 mm

Figure 12. Circumferential Temperature Distribution of Channel 4 and Channel 1

In order to clearly observe the non-uniformity of temperature distribution, the temperature color scale was adjusted as shown in Figure 13. The axial distribution in each channel still exhibited significant non-uniformity. However, due to the influence of the wire spacers, the temperature on the side of Channel 4 near the assembly box wall was higher than that near the heated surface. Overall, the effect of the wire spacers on temperature distribution was more significant than the influence of the distance between fuel rods or between fuel rods and the assembly box wall.

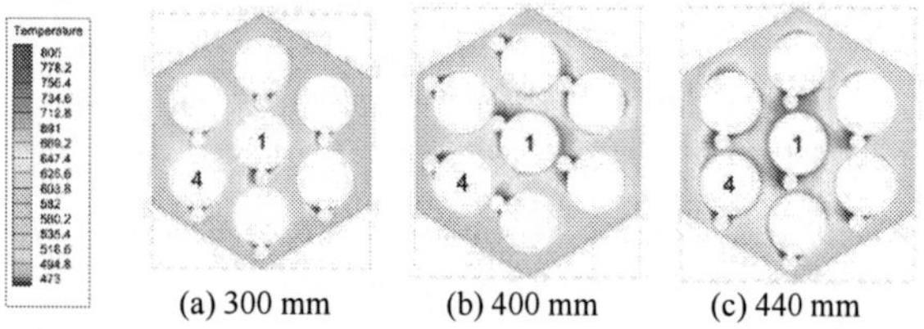

(a) 300 mm   (b) 400 mm   (c) 440 mm

Figure 13. Circumferential Temperature Distribution of Channel 4 and Channel 1 (Legend 2)

V. CONCLUSIONS

This study systematically analyzes the thermal-hydraulic characteristics of a 7-rod fuel assembly under both wire-wrapped and non-wire-wrapped conditions using computational fluid dynamics (CFD) methods. The results demonstrate that for the non-wire-wrapped assembly, the fuel rod surface temperature exhibits a clear axially non-uniform distribution, with the central region showing significantly higher temperatures compared to the sides—a trend directly influenced by the flow characteristics at the inlet section. In contrast, for the wire-wrapped assembly, localized high-temperature zones form at the contact points between the helical wire and the fuel rods. The resulting thermal effects even surpass the impact of the inter-rod gaps on heat transfer, leading to substantial alterations in flow structures, including the observation of anomalous wall temperature distributions in certain regions. The numerical simulation results show good agreement with theoretical analyses, confirming the reliability of the computational model. Furthermore, these findings provide crucial theoretical support and design references for the structural optimization and thermal safety design of compact nuclear fuel assemblies.

REFERENCES

[1] Mikityuk K, Pelloni S, Coddington P, et al. FAST: An advanced code system for fast reactor transient analysis[J]. Annals of Nuclear Energy, 2005, 32(15): 1613-1631.

[2] Luo Xiao, Wang Chi, Zou Zeren, et al. Development and application of a multi-physics and multi-scale coupling program for lead-cooled fast reactor[J].Nuclear Science and Techniques,2022,33(02):42-54.

[3] Shen Xiuzhong, Yu Pingan, Yang Xiuzhou, et al. Inherent Safety Analysis for Lead Cooled Fast Breeder Reactor [J]. Nuclear Power Engineering,2002,23(4):75-78.

[4] Xia Fan, Liu Shuyong, Li Taosheng, et al. Thermal-hydraulic and pressure drop characteristics of lead-bismuth–argon two-phase flow under blocked fuel assembly conditions [J]. Nuclear Safety, 2024, 23(01): 33-47.

[5] Pacio J, Daubner M, Fellmoser F, et al. Experimental study of heavy-liquid metal(LBE) flow and heat transfer along a hexagonal 19-rod bundle with wire spacers [J]. Nuclear Engineering and Design, 2016,301:111-127.

[6] Sifan Dong, Jingguo Wei, Weixiang Wang, et al. Development and application of three-dimensional multi-physics and multi-scale coupling program for lead-cooled fast reactor [J]. Annals of Nuclear Energy, 2025, 219:111486.

[7] Pacio J., Daubner M., Fellmoser F., et al. Experimental study of heavy-liquid metal (LBE) flow and heat transfer along a hexagonal 19-rod bundle with wire spacers [J]. Nuclear Engineering and Design, 2016, 301: 111-127.

[8] Pacio, J., Wetzel, T., Doolaard, H., et al. Thermal-hydraulic study of the LBE-cooled fuel assembly in the MYRRHA reactor: Experiments and simulations. Nuclear Engineering and Design,2017, 312, 327–337.

Hybrid Solution Method for Parallel Multi-Channel Flow Allocation Using Quasi-Newton Method and Quantum Genetic Algorithm

Liu Tao
Suzhou Nuclear Power Research Institute,
Suzhou, Jiangsu Province, China
e-mail: liut@cgnpc.com.cn

Ying Hong
Suzhou Nuclear Power Research Institute,
Suzhou, Jiangsu Province, China
e-mail: yinghong94@126.com

Hui Tianyu*
Suzhou Nuclear Power Research Institute,
Suzhou, Jiangsu Province, China
* Corresponding author's email: 2417245892@qq.com

Wang Qiang
Suzhou Nuclear Power Research Institute,
Suzhou, Jiangsu Province, China
e-mail: 3947646786@qq.com

Zhang Tao
Suzhou Nuclear Power Research Institute,
Suzhou, Jiangsu Province, China
e-mail: 790504071@qq.com

Shi Haining
Suzhou Nuclear Power Research Institute,
Suzhou, Jiangsu Province, China
e-mail: 1107641211@qq.com

Fang Kewei
Suzhou Nuclear Power Research Institute,
Suzhou, Jiangsu Province, China
e-mail: 67350207@qq.com

Abstract—**Parallel multi-channel structures are extensively applied in nuclear engineering and petrochemical industries, with flow distribution posing a critical challenge in fuel assembly reactor modeling. Conventional numerical optimization methods such as quasi-Newton algorithms demonstrate robust convergence but are constrained by initial-value dependencies and computational inefficiency for complex engineering functions. In contrast, modern intelligent algorithms like genetic algorithms (GA) exhibit global search capabilities yet frequently stagnate in local optima. To address these limitations, this paper proposes a hybrid framework combining the Broyden-Fletcher-Goldfarb-Shanno (BFGS) quasi-Newton method and a quantum genetic algorithm (QGA). The hybrid method leverages QGA's global optimization strength to generate high-accuracy initial solutions, thereby mitigating BFGS's sensitivity to initial guesses. Furthermore, a compact QGA population is embedded within the BFGS step-size optimization phase to circumvent gradient calculations for intricate functions and directly identify near-optimal step sizes. Validation on a three-parallel-channel flow distribution problem revealed that the hybrid algorithm converges to the global optimum (optimal value: 0.033131) within 49 iterations, outperforming standalone QGA, which remained trapped in a suboptimal solution (optimal value: 3715.08) after 1,000 iterations. The results underscore the hybrid algorithm's synergistic mechanism: QGA rapidly identifies promising solution regions, while BFGS refines local convergence precision. This dual-mode optimization framework significantly enhances computational efficiency and reliability for complex engineering systems characterized by non-convexity and high dimensionality.**

Keywords- BFGS quasi-Newton method; QGA; Parallel multi-channel structures; Hybrid solution method

I. INTRODUCTION

Parallel channel structures are extensively utilized in critical engineering fields such as nuclear power and electrical systems. In reactor designs incorporating box-type fuel assemblies, flow distribution has emerged as a pivotal research challenge[1]. Traditional approaches rely on the equal-pressure assumption at channel inlets and outlets, solving the nonlinear relationship between pressure drop and flow rate through iterative methods. While precise solutions can be obtained for single-channel scenarios, these methods suffer from exponentially increasing computational complexity in multi-channel systems, significantly diminishing their engineering applicability. The inherent limitations of these conventional methods necessitate a more efficient optimization approach that can handle complex multi-physics interactions while maintaining computational tractability.

Conventional numerical optimization methods, such as quasi-Newton algorithms, demonstrate notable advantages in engineering applications due to their rapid local convergence. However, their heavy reliance on initial guess selection often leads to suboptimal solutions in multimodal problems, while gradient computations in high-dimensional spaces introduce numerical instability. On the other hand, intelligent optimization techniques like GAs eliminate gradient dependencies through stochastic exploration, yet their convergence remains sluggish in fine-tuning phases, and their performance heavily depends on empirical parameter settings such as population size and mutation rates[2][3]. This dichotomy between global exploration and local exploitation highlights a critical gap in current optimization strategies for nuclear engineering applications.

To bridge this gap, this study proposes a hybrid optimization framework that synergizes quasi-Newton algorithms with QGA. The proposed strategy leverages QGA's global exploration capability without initial condition constraints to identify high-potential solution regions[4]. Subsequently, the quasi-Newton algorithm employs its superlinear convergence characteristics for precise local searches. This integration mitigates the impact of GA parameter sensitivity on solution accuracy while reducing redundant computational iterations, thereby offering a novel approach for efficient optimization in complex engineering systems.

II. OPTIMIZATION OF MODEL ESTABLISHMENT

In reactor engineering, the thousands of coolant channels in such a cored reactor with canned fuel assemblies are typically considered as a set of closed parallel channels – i.e., mass and momentum exchange between individual coolant channels is neglected. According to the law of energy conservation, the relationship between coolant temperature and power in each channel is expressed by Eq.(1).

$$Q_i = C_p(T) \cdot W_i \left(T_{out_i} - T_{in} \right) \tag{1}$$

Where, W is the mass flow rate (kg·s^{-1}), Q denotes the power (W), the subscript i represents the i-th coolant channel, C_p is the specific heat capacity at constant pressure (J·kg^{-1}·K^{-1}), T_{out} and T_{in} are the outlet and inlet temperatures (K), respectively.

As shown in Figure 1. , for the parallel multi-channel flow distribution problem, both the mass flow rate W_i and the outlet temperature T_{outi} of each channel are unknown parameters. Consequently, Eq.(1) becomes a two-variable equation.

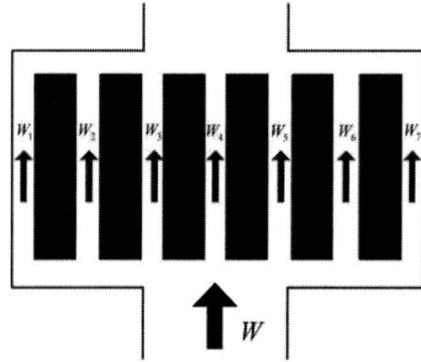

Figure 1.   Schematic Diagram of Parallel Channels

To solve Eq.(1) for determining the flow rates and outlet temperatures of each channel, an initial set of flow rates must be assumed and substituted into Eq.(1) for calculation. The validity of the solution can be verified based on the parallel channel pressure loss equivalence principle, i.e., whether the

pressure drops across all flow channels are equal. The pressure drop loss for each channel is expressed by Eq.(2).

$$\Delta p = \Delta p_g + \Delta p_a + \Delta p_f + \Delta p_c$$

$$= \int_{z_1}^{z_2} \rho_i(z,T)\, g\, dz + \frac{1}{2}\left(\rho_{out} V_{out}^2 - \rho_{in} V_{in}^2 \right) + \left(f \frac{L}{D_e} + \sum K \right)_i \cdot \frac{1}{2} \rho_i \overline{V}_i^2, i=1,2,3.... \tag{2}$$

Where, Dp_g represents gravitational pressure drop (Pa), Dp_a denotes acceleration pressure drop (Pa), Dp_f is frictional pressure drop (Pa), Dp_c and signifies form resistance pressure drop (Pa). V and r are the coolant flow velocity (m/s) and density (kg·m^{-3}), f and K are the Darcy-Weisbach friction factor and local resistance coefficient, respectively. The coolant flow distribution should comply with the principle of equal pressure loss in parallel channels, thus leading to Eq.(3).

$$\Delta P_1 = \Delta P_2 = \Delta P_3 = ... \tag{3}$$

To determine whether the pressure drop within the channels is equal, Eq.(3) can be reformulated into Eq.(4).

$$f = \left(\Delta P_1 - \Delta P_2 \right)^2 + \left(\Delta P_2 - \Delta P_3 \right)^2 + ... \tag{4}$$

The aforementioned equations can be transformed into a constrained optimization problem, as shown in Eq.(5). Given the structure of the problem, the optimal solution is guaranteed to lie within the feasible region.

$$\begin{aligned} \min \quad & f(m_1, m_2, m_3, ...) \\ \text{s.t.} \quad & \sum_{i=1}^{n} m_i = M \\ & m_i > 0, \quad i=1,2,3,.....n \end{aligned} \tag{5}$$

By fixing the first n-1 variables and solving for the n-th variable via the equality constraint, the problem is converted into an unconstrained optimization problem with the transformed objective function Eq.(6) and Eq.(7).

$$m_n = M - \sum_{i=1}^{n-1} m_i \tag{6}$$

$$\min f(m_1, m_2, m_3, ..., m_n) \tag{7}$$

Based on the principle of equal pressure loss in parallel channels, a minimum value of 0 must exist for this problem. However, due to the temperature-dependent properties of the coolant (which are indirectly linked to flow rate), Eq.(2) becomes a highly nonlinear and complex relationship. Assuming the inlet face is an isobaric surface and using NaK (sodium-potassium alloy) as the coolant, the relationship between the pressure drop in a specific channel and its outlet temperature can be derived using Python.

III. DEVELOPMENT OF HYBRID OPTIMIZATION ALGORITHM

To address the initial value sensitivity and derivative calculation limitations of quasi-Newton methods, along with the convergence issues of genetic algorithms, this work integrates the quasi-Newton method with the Quantum Genetic Algorithm (QGA) through a two-phase coupling:

(1) QGA iteratively generates high-quality initial solutions for the quasi-Newton method;

979-8-3315-2490-6/25 $31.00 © 2025 IEEE

(2) During quasi-Newton iterations, a small-population QGA directly optimizes step sizes, bypassing analytical complexities.

The proposed method enables rapid global optimization without predefined initial conditions, reduces sensitivity of solution accuracy to GA parameter selection, enhances convergence speed while maintaining precision, and significantly lowers computational costs.

A. Quasi-Newton Method

Common quasi-Newton methods include the Davidon-Fletcher-Powell (DFP) algorithm[5] and the BFGS algorithm[6][7]. The DFP algorithm is categorized as a rank-2 quasi-Newton algorithm, which combines the advantages of the gradient method and Newton's method while eliminating the need to compute the inverse of the Hessian matrix. The procedural steps of the DFP algorithm are outlined as following.

Step 1: Given an initial value x_0, set the convergence tolerance $e > 0$.

Step 2: Make $H_0 = E_n$, compute $g^0 = \nabla f(x^0)$ and set $k=0$.

Step 3: Make $p^k = -H_k g_k$.

Step 4: Determine the optimal step sizec l_k by Eq.(8).

$$f\left(x^k + \lambda_k p^k\right) = \min_{\lambda>0}\left(x^k + \lambda p^k\right) \tag{8}$$
Set $x^{k+1} = x^k + l_k p^k$, compute $g^{k+1} = \nabla f(x^{k+1})$.

Step 5: If the condition $g^k < e$ is met, stop computation and output the result; otherwise, proceed to Step 6.

Step 6: If $k=n$-1, set $x^0 = x^{k+1}$ and back to Step 2; otherwise, proceed to Step 7.

Step 7: Compute the Eq.(9)~Eq.(11).

$$\Delta g^k = g^{k+1} - g^k \tag{9}$$
$$\Delta x^k = x^{k+1} - x^k \tag{10}$$
$$H_{k+1} = H_k + \Delta g^k = H_k + \frac{\Delta x^k \left(\Delta x^k\right)^T}{\left(\Delta x^k\right)^T \Delta g^k} - \frac{H_k \Delta g^k \left(\Delta g^k\right)^T H_k}{\left(\Delta g^k\right)^T H_k \Delta g^k} \tag{11}$$
Set $k=k+1$ and back to Step3.

The DFP algorithm can rapidly converge to optimal solutions and is widely used in engineering applications. However, it suffers from numerical instability. The BFGS algorithm addresses this limitation by improving the numerical stability of the DFP method. While its computational procedure resembles that of DFP, BFGS effectively serves as the dual formulation of the DFP problem. As shown in Eq.(12), the primary distinction between BFGS and DFP lies in the use of matrix B (in BFGS) instead of matrix H (in DFP).

$$B_{t+1}^{-1} = \left(I_n - \frac{\Delta x_t \Delta g_t^T}{\Delta x_t^T \Delta g_t}\right) B_t^{-1} \left(I_n - \frac{\Delta g_t \Delta x_t^T}{\Delta x_t^T \Delta g_t}\right) + \frac{\Delta x_t \Delta x_t^T}{\Delta x_t^T \Delta g_t} \tag{12}$$

B. Quantum Genetic Algorithm

Genetic algorithms (GAs) simulate biological evolutionary processes, using fitness values as the decision criterion to search for optimal solutions within defined boundaries. The calculation of fitness is particularly critical, as it directly influences the final computational results. Theoretically, the correct values for internal and external outlet temperatures should be determined such that the pressure drops across the internal and external coolant channels are equalized. Conventional genetic algorithms, however, often lack maturity and flexibility. This is because their global and local search capabilities—governed by the crossover and mutation operations—are highly sensitive to parameter settings. Variations in these parameters can lead to divergent computational outcomes.

To address the limitations of conventional GAs, scholars have proposed QGA that integrates concepts from quantum computing, thereby significantly enhancing the global search capabilities of genetic algorithms. Figure 2. and Figure 3. present the flowcharts of both standard genetic algorithms and quantum genetic algorithms.

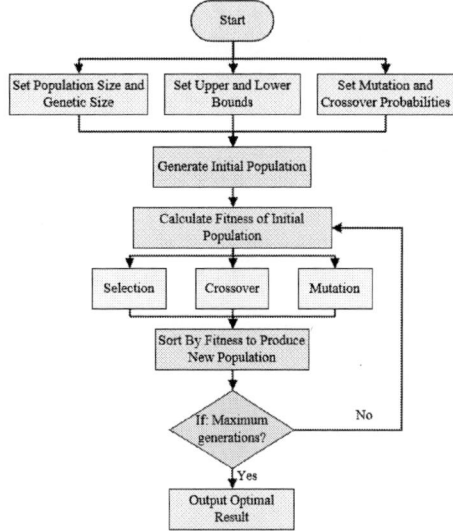

Figure 2.   Standard Genetic Algorithm Structure Diagram

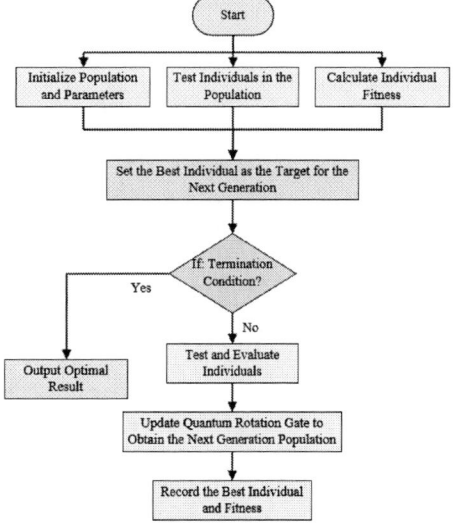

Figure 3.   Quantum Genetic Algorithm Structure Diagram

QGA introduces the following key enhancements over conventional approaches:

1.Quantum-inspired Encoding: Employs quantum state vectors for genetic representation and leverages quantum logic gates for chromosome evolution, yielding superior results compared to standard genetic algorithms.

2.Probabilistic Superposition Encoding: Utilizes qubit probability amplitudes for chromosome encoding, enabling a single chromosome to represent a superposition of multiple states.

C. Hybrid Algorithm

To address the challenges of initial value sensitivity and gradient dependency in quasi-Newton methods, as well as the convergence issues in genetic algorithms, this study integrates BFGS/DFP algorithms with the QGA. The computational workflow is illustrated in Figure 4.

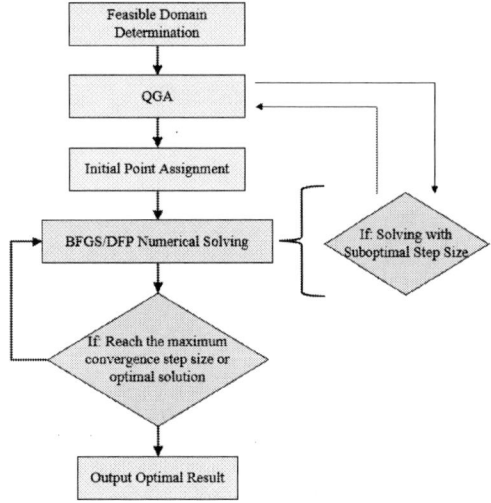

Figure 4. Hybrid Algorithm Structure Diagram

According to the flowchart shown in Figure 4. , the coupling strategy consists of two integrated components. The QGA exhibits strong global search capability, enabling it to efficiently explore the entire solution space and identify optimal solutions. This allows the QGA to iteratively generate high-quality initial solutions for the quasi-Newton method, effectively addressing the initialization sensitivity issue inherent in quasi-Newton algorithms. Furthermore, when determining the optimal step size, the quasi-Newton method incorporates a small-population QGA for direct optimization, which bypasses the computational challenges of complex function evaluations while leveraging the QGA's robust global exploration ability to obtain near-optimal step lengths within just a few dozen iterations. Through this synergistic integration of the two approaches, the hybrid algorithm achieves computationally efficient and reliable optimization performance.

IV. SOLUTION FOR PARALLEL THREE-CHANNEL PROBLEM

A. Problem Statement

This paper selects a parallel three-channel system as the research object, as shown in Figure 5. Assuming the total mass flow rate M, the fluid flows from a single channel into three channels, being divided into three branches with flow rates denoted as m_1, m_2, and m_3, respectively. If the inlet plane is considered an equipressure surface, then $\Delta P_1 = \Delta P_2 = \Delta P_3$, leading to the following objective function.

$$\min \quad \left(\Delta P_1 - \Delta P_2\right)^2 + \left(\Delta P_2 - \Delta P_3\right)^2 \tag{13}$$

Figure 5. Parallel Three-Channel Problem System

Based on the law of equal pressure drop in parallel channels, Eq.(6) must have an optimal solution of zero. The geometric and thermal-hydraulic parameters of the channels are listed in TABLE I.

TABLE I. PARAMETERS OF PARALLEL THREE-CHANNEL PROBLEM SYSTEM

Parameters	Value	Unit
Kind of Fluid	NaK Alloy	/
Total Mass Flow Rate	3	kg/s
Inlet Temperature	450	K
Length of Channel	2	m
Diameter of Channel	0.05	m
Heat in Channel 1	2×10^4	W
Heat in Channel 2	1×10^4	W
Heat in Channel 3	0.5×10^4	W

For simplified calculation, this work takes the outlet temperatures T_1, T_2 as independent variables. Based on Eq.(1), the mass flow rates m_1, m_2 in Channel 1 and Channel 2 can be derived. Then, the mass flow rate m_3 in Channel 3 can be obtained from Eq.(5). Subsequently, the pressure drops ΔP_1, ΔP_2, ΔP_3 across each channel are calculated. This leads to the formulation of the objective function expression

B. Results and Discussion

In the computation, the parameters for the QGA algorithm were set as shown in 0In the BFGS algorithm, convergence was considered achieved when

$$\min \quad \left(\Delta P_1 - \Delta P_2\right)^2 + \left(\Delta P_2 - \Delta P_3\right)^2 < 1 \tag{14}$$

TABLE II. PARAMETER SETTINGS OF QGA

	Parameters	Value
Initial Iteration QGA	Population	10
	Chromosomes number	2
	Chromosome length	20
	Number of iterations	40
	Initial range	[450,600]
	Quantum rotation angle	0.3141
Get Optimal Step Size QGA	Population	10
	Chromosomes number	2
	Chromosome length	20
	Number of iterations	25
	Initial range	[0,2]
	Quantum rotation angle	0.31415926
Individual QGA Calculation	Population	10
	Chromosomes number	2
	Chromosome length	20
	Number of iterations	1000
	Initial range	[450,600]
	Quantum rotation angle	0.31415926

Figure 6 presents the computational results of the hybrid optimization algorithm versus the standalone quantum genetic algorithm (QGA). The hybrid algorithm found the optimal solution after 49 iterations, with the outlet temperatures of channels 1/2/3 being 470.918 K, 460.941 K, and 455.603 K, respectively, and corresponding mass flow rates of 1.0403 kg/s, 0.9929 kg/s, and 0.9688 kg/s. The optimal objective value was 0.033131. In comparison, the QGA required 1,000 iterations to reach its best solution, with outlet temperatures of 472.386 K, 461.256 K, and 455.104 K, yielding an optimal value of 3715.08. The results demonstrate that the hybrid algorithm developed in this work achieves the optimal solution in significantly fewer iterations, while the standalone QGA fails to converge to the true optimum even after 1,000 iterations. As shown in Figure 6, the QGA exhibits strong global search capability, locating the vicinity of the optimal solution within just a dozen iterations. However, its poor local search ability causes it to remain trapped in a suboptimal region for an extended period.

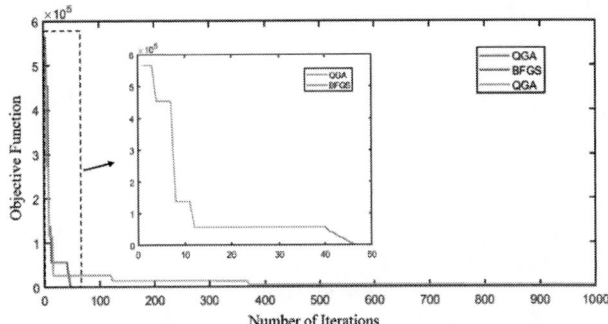

Figure 6. Performance Comparison: Hybrid Algorithm vs. Quantum Genetic Algorithm

Figure 7 presents the calculated objective function values near the optimal solution. A steep decline (from 105 to nearly 0) is observed in the vicinity of the minimum, indicating a stringent requirement for high local search precision. While the quantum genetic algorithm (QGA) demonstrates strong global search capability, it struggles with local exploitation, often stagnating in suboptimal regions. In contrast, the hybrid algorithm developed in this work combines QGA's global exploration with the BFGS algorithm's powerful local convergence properties. Once QGA locates the approximate vicinity of the optimum, BFGS rapidly refines the solution within just a few iterations, significantly reducing computational overhead and improving efficiency. This hybrid approach proves particularly effective for parallel-channel optimization problems, offering both computational economy and shorter convergence times, making it highly applicable in engineering contexts.

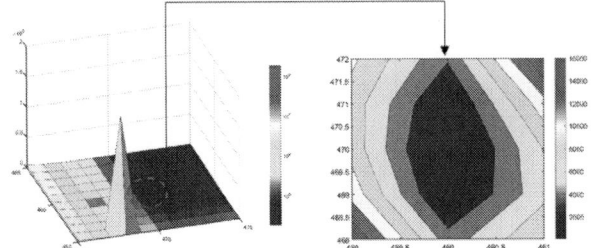

Figure 7. Objective function values near the minimum

V. CONCLUSION

In this study, a novel hybrid optimization algorithm is proposed by integrating the Broyden–Fletcher–Goldfarb–Shanno (BFGS) method, which exhibits strong local search capability, with a quantum genetic algorithm (QGA) that offers superior global exploration. Using parallel channel optimization as a case study, the performance of the proposed hybrid algorithm was evaluated. Comparative results demonstrate that, in contrast to standalone QGA, the hybrid algorithm efficiently locates the vicinity of the optimal solution in the global search space and subsequently achieves rapid convergence to the exact optimum within a few iterations, successfully combining robust global exploration with precise local refinement. The developed algorithm exhibits high computational efficiency in parallel channel optimization, significantly reducing optimization time and computational costs, thereby demonstrating clear practical value in engineering applications. Future research could further extend this hybrid approach to solving other complex optimization problems, providing a new and effective computational tool for engineering applications.

REFERENCES

[1] Li Lei, Zhang Zhijian, Research on Transient Flux Distribution in Parallel Channels, Nuclear Power Engineering, 2010,31(05):97-101.

[2] Wang Xianbin, Shi Shuming, Liu Li, Jin Lisheng, Hybrid solution method for vehicle dynamics equilibrium points based on genetic algorithm and quasi-Newton method, Journal of Mechanical Engineering, 2014, 50(4): 120-127.

[3] Arute F, Arya K, Babbush R, et al. Quantum supremacy using a programmable superconducting processor[J]. Nature, 2019, 574(7779): 505-510.

[4] Xu Lianming, Wang Xiaojun, Wang Zhenghuan, Cao Geyong. Hybrid quantum genetic algorithm for structural damage identification[J]. Computer Methods in Applied Mechanics and Engineering, 2025, 438(Part B): 117866.

[5] Zhou Weijun. A modified BFGS type quasi-Newton method with line search for symmetric nonlinear equations problems [J]. Journal of Computational and Applied Mathematics, 2020, 367: 112454.

[6] Yabe Hiroshi, Ogasawara Hideho, Yoshino Masayuki. Local and superlinear convergence of quasi-Newton methods based on modified secant conditions [J]. Journal of Computational and Applied Mathematics, 2007, 205(1): 617–632.

[7] Wenjuan X ,Chungen S ,Zhensheng Y .Limited Memory BFGS Method for Least Squares Semidefinite Programming with Banded Structure[J].Journal of Systems Science & Complexity,2022,35(04):1500-1519.

2025 International Conference on Advanced Energy Systems and Power Electronics (AESPE 2025)

Evaluation of input and output of costly project resources in power grid enterprises System Research

Guocheng Li[1]

State Grid Shandong Electric
Power Company Dezhou Power
Supply Company
@dz.sd.sgcc.com.cn

Yuansheng Duan[2]

State Grid Shandong Electric
Power Company
@sd.sgcc.com.cn

Xiaoxue Ma[3,*]

State Grid Shandong Electric
Power Company Dezhou Power
Supply Company
2243041393@qq.com

Can Tao[4]

State Grid Shandong Electric
Power Company Dezhou Power
Supply Company
475420353@qq.com

Abstract: **In order to comply with the development requirements of precise input and quality enhancement, and to closely integrate the development strategy and financial priorities of power grid enterprises, this paper firstly builds a linkage index system of resource input and output to pave the way for quantifying the effect of resource input; secondly, it builds a multidimensional evaluation system of resource input and output, and arrives at the scores of each dimension of the costly projects; finally, it uses the hierarchical analysis method to give the corresponding weight to the different dimensions, and arrives at the final scores, which will provide the basis of future resource allocation of power grid enterprises, and thus promote the development and progress of the distribution network. Finally, the hierarchical analysis method is used to assign corresponding weights to different dimensions, and the final score is obtained, which provides the basis for the resource allocation of the future power grid enterprises, and then promotes the development and progress of power distribution networks.**

Keywords: power grid companies, cost-based projects, resource input-output evaluation

I. INTRODUCTION

At present, the international community intensified trade protectionism, geopolitical risks increase, the domestic economy rebound to a good foundation is not yet stable, the new and old momentum transformation, new energy development on the company's transmission effect gradually highlights the construction of a new type of electric power system into a critical period of time, the complexity of the changing internal and external environments, on the grid enterprises to enhance the ability to integrate resources, enhance the quality of inputs and outputs of the efficiency of the proposed higher requirements.

Chen Yanli et al[1] sorted out the problems existing in the incremental budget management mode of power grid enterprises, and in view of the problems, put forward a new model of budget management that integrates zero-based budgeting and operation-based budgeting. This model realizes the standardization of project feasibility study audit decision-making and online audit standardization management through the cost project library management platform. Li Panpan et al[2] analyzed the key factors affecting the cost management of power grid enterprises, and on this basis, pointed out the main problems existing in the cost management of power grid enterprises and put forward optimization strategies. The research of Gao Siqin et al[3] provides new perspectives and methods for the cost risk control of distribution network project

materials in power grid enterprises, especially in the construction of informationization and performance evaluation system. Zhou Yunhan[4] mainly discusses the practical methods and problems of cost control in project financial management of power grid enterprises, and puts forward improvement strategies. Wang Shanshan[5] explored the cost management problems and strategies of power grid enterprises in the context of multidimensional lean management change in his study, and proposed strategies to improve the status quo of cost management through the establishment of a sound cost management guarantee system, accurate refinement of the cost management data and increasing the use of multidimensional data to achieve the optimal allocation of resources and maximize the benefits of the enterprise, so as to support the enterprise to achieve the fierce market competition in the sustainable development. In his study, He Pusheng[6] explored the impact of transmission and distribution price reform on the cost management of power grid enterprises, and put forward strategies such as whole-life cost management of grid assets, implementation of the balancing account system, and implementation of lean cost management, with a view to improving the level of cost management and market competitiveness of power grid enterprises, and to ensure the sustainable development of enterprises.

To summarize, most of the current scholars for the grid enterprise cost research is qualitative, for how to build a scientific and reasonable performance evaluation system is not deep enough, and did not study the long-term effect of cost management strategies and sustainability of the impact. In this paper, we build a multi-dimensional resource input and output evaluation system to quantitatively evaluate the long-term input effect of resources and provide a more scientific and reasonable basis for power grid enterprises to guide them to better manage resource allocation.

II. RESEARCH IDEAS

This paper evaluates costly project management and input-output effects by integrating hierarchical analysis and multi-dimensional operation management. We select indicators for resource investment in quality, efficiency, and management maturity. An evaluation system with three dimensions—unit resource input effect, efficiency, and goal achievement—is established with a scoring mechanism. Using hierarchical analysis aligned with grid enterprise development, we assign weights to these dimensions and calculate each project's comprehensive score, as illustrated in Figure 1.

979-8-3315-2490-6/25 $31.00 © 2025 IEEE

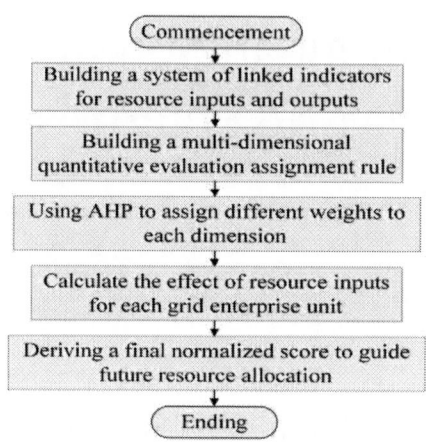

Figure 1 Diagram of the research idea

III. INVESTMENT OUTPUT EVALUATION MODEL AND APPLICATION OF RESULTS

This paper first clarifies the evaluation objective, which is to quantify the resource input and output effects of cost-based programs in power grid enterprises. To this end, this paper designs an evaluation system containing three dimensions (unit resource input effectiveness, unit resource input efficiency, and degree of expected goal realization). Each dimension is measured by specific indicators, such as the percentage of decrease in distribution fault index. The experimental design includes data collection, calculation of indicators, calculation of dimension scores and comprehensive evaluation scores.

A. System of resource input-output indicators

The project targets marketing and equipment professions, identifying key improvement indicators based on resource investment and expected outcomes. It forms a system linking resource investment and business enhancement layers. Indicators with high relevance and low correlation with the rest of the project should be selected. In addition, this paper also refers to the business enhancement needs of grid enterprises and the requirements of the State Grid Headquarters to ensure the practicality and scientificity of the indicator selection. In the resource investment layer, equipment includes 6 primary and 41 secondary indicators like grid risk management and emergency repairs; marketing includes 6 primary indicators like metering and power service, plus 41 secondary ones. In the

business enhancement layer, equipment covers indicators like power supply reliability and grid strength, with secondary ones such as line outage rates; marketing covers service quality and green development, with secondary indicators like billing accuracy and customer satisfaction.

B. Three-dimensional scoring rules and points

The evaluation model in this paper collects resource inputs (funds, personnel, etc.), output index data (distribution failure index, etc.), and expected resource input targets of cost projects in the past three years, and the data should be true and accurate, and the collected data are used to calculate the proportion of output indexes that are improved or decreased by the unit of inputs (hereinafter referred to as the "effect of the unit of resource inputs"), and measure the effect of each unit of resource inputs and outputs from three dimensions. The unit resource input and output effects are measured in three dimensions, and the scoring rules for each dimension are as follows.

(1) Evaluation dimension of unit commissioning effect

This dimension is designed in the evaluation system in order to discern the differences in the resource input and output effects of each power supply unit. This dimension ranks each power supply unit by calculating the average value of the unit resource input effect of each unit in the last three years, and assigns points to this dimension according to the rules in Table 1.

Table 1 Rules for assigning points to the effect of unit production

serial number	rankings	score
1	1-3	100
2	4-7	90
3	8-11	80

(2) Evaluation dimension of unit commissioning efficiency

In order to reasonably evaluate the resource input and output efficiency of each power supply unit, this dimension is designed in the evaluation system. This dimension divides each power supply unit into three categories of growing, fluctuating

and declining according to the change trend of its unit resource input effect, and scores them according to the rules in Table 2.

979-8-3315-2490-6/25 $31.00 © 2025 IEEE

Table 2 Rules for assigning points for unit production efficiency

serial number	classifications	score
1	Growth type (liters)	100
2	Fluctuating type (descending or ascending)	90
3	Descending type (descending)	80

(3) Evaluation dimension of the degree of realization of expected goals

In order to motivate the power supply units to make efforts to realize the expected goals in the future, this dimension is designed in the evaluation system. This dimension compares the expected targets with the final actual output indicators, divides the power supply units that have achieved the expected targets and those that have not, and assigns points according to the rules in Table 3.

Table 3 Scoring rules for the degree of achievement of expected objectives

serial number	classifications	score
1	Completion of expected target units	100
2	Units not meeting expected targets	80

C. Three-dimensional weighting design based on AHP

Analysis of Hierarchy (AHP) is a method of system analysis and decision-making, proposed by American operations researcher Satie in the 1970s. It decomposes a complex problem into levels of objectives, criteria and solutions by establishing a hierarchical model, then compares the elements of each level in pairs, constructs a judgment matrix and calculates the weights, and finally synthesizes the decision-making results. This method combines qualitative and quantitative analysis, can effectively deal with multi-criteria decision-making problems, and is widely used in the fields of economic management, engineering technology, social decision-making, etc. It helps decision-makers weigh the pros and cons of multiple factors and make scientific and reasonable decisions.

The AHP calculates the subjective weight coefficients of each evaluation index by applying the geometric mean method based on the constructed judgment matrix, which is calculated as follows:

$$\begin{cases} w_{ki} = \dfrac{\overline{w_{ki}}}{\sum\limits_{i=1}^{n} \overline{w_{ki}}} = \dfrac{\sqrt[n]{\prod\limits_{j=1}^{n} d_{ij}}}{\sum\limits_{i=1}^{n} \overline{w_{ki}}} \\ w_{sj} = w_k \times w_{ki} \end{cases} \quad (1)$$

Where: w_{ki} is the weight coefficient of the ith evaluation index under the kth criterion, n is the number of evaluation indexes, w_k is the weight coefficient of the kth criterion, and w_{sj} is the subjective weight coefficient of the jth key feature quantity.

In order to judge whether the results of hierarchical analysis

method are scientific and reasonable, the consistency test is carried out by calculating the consistency index (CI) and consistency ratio (CR). If CR < 0.1, the consistency of the judgment matrix is acceptable, otherwise the judgment matrix needs to be adjusted until it meets the requirements. The specific formulas for CI and CR are as follows:

$$\begin{cases} CI = \dfrac{(\lambda_{\max} - n)}{(n-1)} \\ CR = \dfrac{CI}{RI} \end{cases} \quad (2)$$

Where: $\lambda_{\max}$ is the largest characteristic root of the judgment matrix, n is the matrix order, and RI is the average random consistency index.

This paper uses the hierarchical analysis method to assign weights to three dimensions in an evaluation model for power grid enterprises' future development, in line with State Grid Headquarters' requirements. A questionnaire gathers opinions from grid enterprise leaders and experts to form a judgment matrix integrating target and index layers. Normalized calculations determine the weights: unit commissioning effect (50%), unit commissioning efficiency (30%), and expected goal achievement (20%), making grid enterprise rankings more intuitive.

IV. EMPIRICAL ANALYSIS

This paper selects the 10kV distribution line maintenance cast for case analysis, which casts the resource input and output index for distribution failure index. This case collects A, B company and other 11 Shandong Province power grid enterprises in 2021, 2022, 2023 three years, according to each data calculation of each power grid enterprise unit resource input effect as shown in Table 4.

Table 4 Effectiveness of resource inputs by grid enterprise units

serial number	statistical unit	Proportion of decrease in distribution fault index per unit of input			Three-year average
		2021	2022	2023	

1	A	-1.31%	0.02%	0.10%	-0.40%
2	B	0.05%	0.22%	0.53%	0.26%
3	C	0.12%	0.00%	0.06%	0.06%
4	D	0.13%	0.07%	0.09%	0.10%
5	E	0.31%	0.06%	-2.33%	-0.65%
6	F	0.11%	0.06%	0.03%	0.07%
7	G	0.15%	0.00%	0.01%	0.05%
8	H	0.20%	0.04%	0.07%	0.10%
9	I	0.48%	0.78%	-1.41%	-0.05%
10	J	0.09%	0.15%	-0.20%	0.02%
11	K	0.12%	0.83%	0.00%	0.32%

According to the unit resource input effect of each grid enterprise in the past three years, considering the three-dimension assignment rules and taking into account the different weights of the three dimensions, the final score of each company is derived as shown in Figure 2.

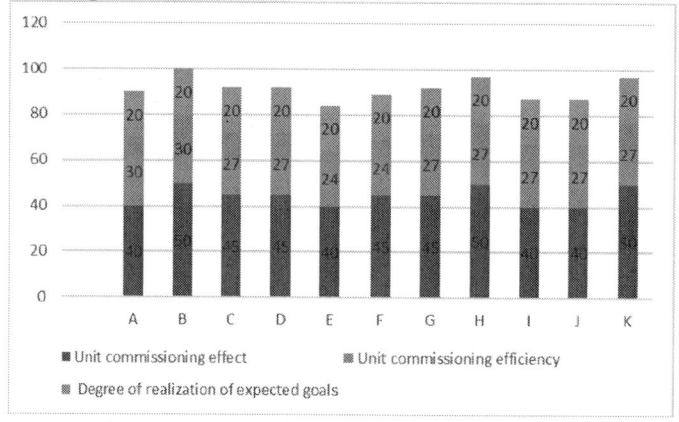

Figure. 2 Final scores of three-dimensional quantitative evaluation of each company

According to the resource input and business index enhancement of each grid company, using the quantitative evaluation method constructed in this paper, the resource input effect of each grid company can finally be evaluated quantitatively, so as to provide a scientific basis for the optimization of resource allocation in the future of the grid enterprise, and to enhance the ability of resource integration.

In order to verify the reliability of the method, this paper conducted a consistency test. By calculating the consistency index (CI) and consistency ratio (CR) of the judgment matrix, it ensures that the CR value is less than 0.1, thus confirming that the consistency of the judgment matrix is acceptable. In addition, this paper further verifies the validity of the methodology by empirically analyzing and comparing the evaluation results with the actual business improvement.

V. CONCLUSION

The objective of this study is to construct a multi-dimensional evaluation model to provide a scientific basis for the resource allocation of power grid enterprises. The realization of the study is based on the Analytic Hierarchy Process (AHP) and multi-dimensional management theory, and the effectiveness of the model is verified through empirical analysis. Compared to existing research, the innovation of this study lies in providing a quantitative method to evaluate the effectiveness of resource input and output, which not only enriches the quantitative research on cost management of power grid enterprises but also offers new perspectives and methods for enhancing resource coordination capabilities and improving input-output efficiency under complex and changing internal and external environments. Additionally, this study points out the limitations of the model, namely, the lack of consideration of the intercorrelation among output indicators, providing directions for future research improvements.

REFERENCE

[1] CHEN Yanli, BU Xianglei, MA Xing, et al. Research on the innovation mode of cost and expense management in power grid enterprises-- Integration and application of zero-based budget and operation-based budget[J]. Finance and accounting newsletter,2025,(04):165-170.

[2] Li Panpan. Exploration on cost management optimization strategy of power grid enterprises[J]. Quality and Market, 2024,(11):105-107.

[3] Gao Sichin, Wang Xintao, Lu Chun. The thinking of material cost risk control of distribution network project of power grid enterprise[J]. Finance and Accounting,2025,(02):71-72.

[4] Zhou Yunhan. Cost control work practice in project financial management of power grid enterprises[J]. China market,2024,(36):153-156.

[5] Wang Shanshan. Research on cost management of power grid enterprises based on multidimensional lean management change[J]. China Management Informatization,2025,28(02):31-33.

[6] He Pusheng. Cost management problems and improvement strategy of power grid enterprises under the background of transmission and distribution price reform[J]. Wisdom China,2024,(12):33-35.

Research on Energy Storage Mechanism of Polymer Composite Materials Based on Interface Structure Regulation

Chen Chen[*1,2], Lifang Shen[1,2], Shubin Yan[1,2], Guang Liu[1,2] and Yang Cui[1,2]

[1]School of Electrical Engineering, Zhejiang University of Water Resources and Electric Power, Hangzhou, China

[2]Zhejiang-Belarus Joint Laboratory of Intelligent Equipment and System for Water Conservancy and Hydropower Safety Monitoring, Hangzhou 310018, China

[*]cc@zjweu.edu.cn

Abstract-Thin film capacitors play an indispensable role in the field of green energy. Thin film capacitors have the characteristics of high voltage resistance, good temperature characteristics, and stable performance, which make them widely used in new energy technologies such as solar photovoltaic power generation systems and wind power generation systems. With the rapid development of polymer film capacitors, their energy loss or efficiency is receiving increasing attention in practical applications. Lower energy storage efficiency means a significant loss of electrical energy. Research has shown that using magnetron sputtering to deposit boron nitride (BN) inorganic layers with good insulation properties on both sides of linear dielectric polyethylene naphthalate (PEN), BN/PEN/BN multilayer composite films have achieved energy storage performance (6.10J/cm³) at corresponding breakdown strengths (500MV/m). The presence of inorganic layers on the surface can suppress the introduction of electrode charges, thereby reducing energy loss.

Keywords- Interface Structure Regulation, Energy Storage Mechanism, Polymer Composite Materials

I. INTRODUCTION

In the field of mechatronics engineering, thin film capacitors also have a wide range of applications and important contributions. Thin film capacitors have been widely used in electronic devices due to their small size, light weight, and high reliability. In mechatronics engineering, thin film capacitors can be used for power regulation, energy recovery, electromagnetic compatibility (EMC) suppression, and DC filtering. For example, in new energy vehicles, thin-film capacitors are used for power regulation, which can effectively extend battery life, improve vehicle safety performance and driving efficiency. Meanwhile, in the process of energy recovery and storage in new energy vehicles, thin-film capacitors work together with battery packs to convert the kinetic energy generated during the braking process into electrical energy and store it. In addition, thin film capacitors can also have an EMC suppression effect, reducing the harm of electromagnetic radiation to the surrounding environment and human body. Dielectric capacitors have many performance advantages and are used in the fields of inverters and pulse power supplies. Therefore, research on the energy storage density of related polymers has begun to receive

attention, and corresponding research results are constantly increasing[1-3].

In recent years, energy storage dielectrics have developed rapidly and achieved many results. With the deepening of research, the focus of energy storage research has also undergone a series of changes. Firstly, inorganic nanofillers can better improve the electrical properties of polymer matrices, but their mechanical strength will decrease due to the difference in dielectric constants between the two. The decrease in mechanical strength of polymers hinders the overall excellent energy storage performance of composite materials[4,5].

Previous studies have found that electrical properties can affect the energy of materials to some extent, but the increase in energy density is also influenced by filling thresholds and performance differences. Finding the optimal preparation process for polymer based composite materials is of great significance for the performance of composite materials. Therefore, researchers considered synergistically improving the dielectric constant and breakdown strength by constructing different composite materials into topological structures with interlayer structures[6-9].

It is necessary to appropriately improve the charging and discharging efficiency for such materials. The ratio of stored electrical energy to released electrical energy during the discharge process. Therefore, low charging and discharging efficiency will result in a significant amount of energy waste. Therefore, it is necessary to address the composition of dielectric loss in order to improve efficiency. There are two main mechanisms of dielectric loss, namely conductivity loss and polarization loss. The problems of interface corrosion and nanoparticle aggregation can increase the carrier density, leading to high dielectric loss. It can be inferred that reducing carrier mobility is important to improve the efficiency of this material[10].

Using magnetron sputtering technology to prepare inorganic functional layer/energy storage medium/inorganic functional layer, where the inorganic functional layer is BN and the energy storage medium is selected as linear medium PEN. Introducing an inorganic BN layer with good insulation properties can suppress electrode charge injection. The prepared multilayer composite film exhibits excellent energy storage performance under appropriate electric field strength.

II. EXPERIMENTAL SECTION

A. Preparation of BN/PEN/BN Multilayer Films

As shown in Figure 1, Using magnetron sputtering equipment to deposit a certain thickness of inorganic layers on a given original thin film by adjusting gas flow rate and sputtering pressure, controlling the different positions of the inorganic layer relative to the polymer layer to obtain multilayer thin films with different structures.

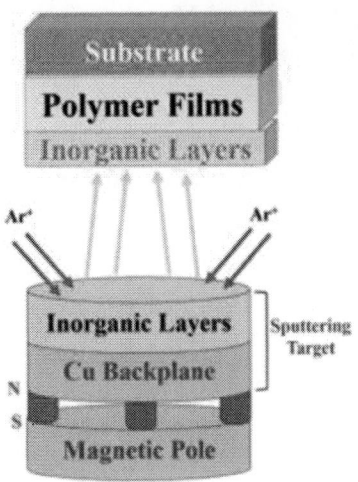

Figure 1. Preparation of BN/PEN/BN Multilayer Films.

B. Characterization and Performance Testing

The phase composition of the film at 40kV and 40mA was detected using X-ray diffraction (XRD) and Cu K_α. The surface and cross-sectional morphology of the composite material were observed using field emission scanning electron microscopy (FESEM). To measure dielectric parameters, aluminum electrodes (diameter 25mm) were deposited on both sides of the composite material, and a broadband impedance analyzer was used to obtain the relative dielectric constant and dielectric loss tangent values in the frequency range of 10Hz to 1M Hz at room temperature. For energy storage performance testing, a ferroelectric testing system is used to test the polarization curve of the sample, and then the energy storage density and charge discharge efficiency are calculated.

III. RESULTS AND DISCUSSION

A. Microstructure of BN/PEN/BN Multilayer Films

As shown in the XRD pattern of the BN/PEN/BN multilayer film in Figure 2, the broad peak near $2\theta=28°$ represents the presence of polymer PEN. Due to the amorphous nature of AlN, no diffraction peaks were observed. Figure 3 shows the microstructure of the composite material, with an AlN insulation layer of approximately 100nm.

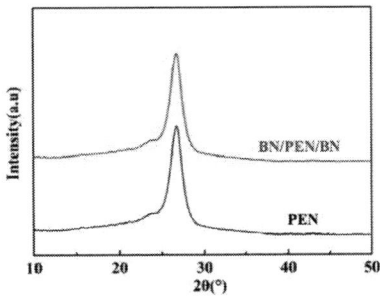

Figure 2. XRD pattern of BN/PEN/BN multilayer film.

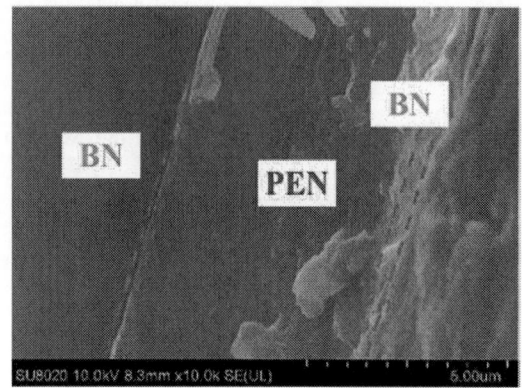

Figure 3. SEM image of BN/PEN/BN multilayer film.

B. Electrical Properties of BN/PEN/BN Multilayer Films

According to the dielectric test results shown in Figure 4, the dielectric constant of BN/PEN/BN film at 100 ℃ is 3.62, which is 1.02 times that of pure PEN (~3.53), possibly due to interface polarization.

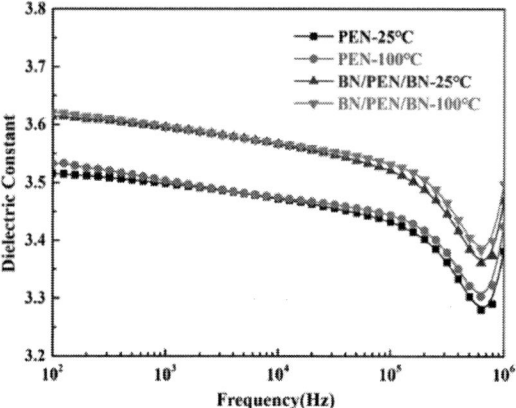

Figure 4. Dielectric constant of BN/PEN/BN multilayer film.

In Figure 5, the dielectric loss of the thin film decreases with the introduction of the insulating layer, because when BN acts as the surface insulating layer in contact with the electrode, it hinders the injection of charges.

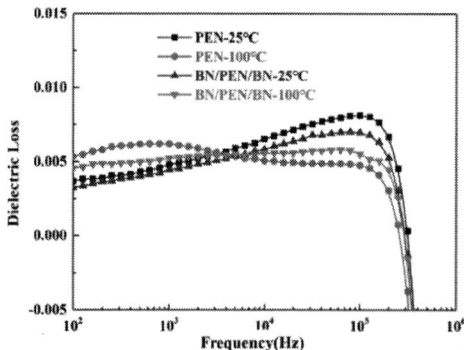

Figure 5. Dielectric Loss of BN/PEN/BN Multilayer Films.

C. *Energy storage performance of BN/PEN/BN multilayer films*

The polarization curve shown in Figure 6 can be used to study the polarization behavior of multilayer films. It can be observed that the introduction of the BN surface insulation layer reduces the residual polarization of the multilayer film, and the smaller the residual polarization value, the lower the loss of the multilayer film.

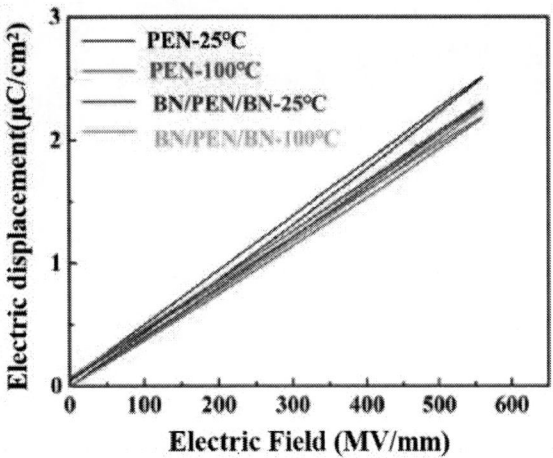

Figure 6. Polarization curves of BN/PEN/BN multilayer films.

After charging and discharging the composite film under different electric field strengths in the specific temperature environment shown in Figure 7. At a temperature of 100 °C, the energy density of the multilayer medium is 6.10J/cm³.

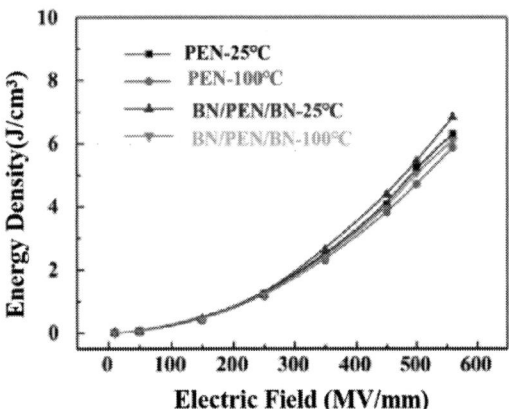

Figure 7. Discharge energy density of BN/PEN/BN multilayer film.

Meanwhile, by integrating and comparing the polarization curves, the efficiency of the dielectric can be obtained. As shown in Figure 8, the efficiency can reach over 90% in different states. This is also because the presence of BN insulation layer can improve the overall breakdown strength and leakage of the dielectric.

979-8-3315-2490-6/25 $31.00 © 2025 IEEE

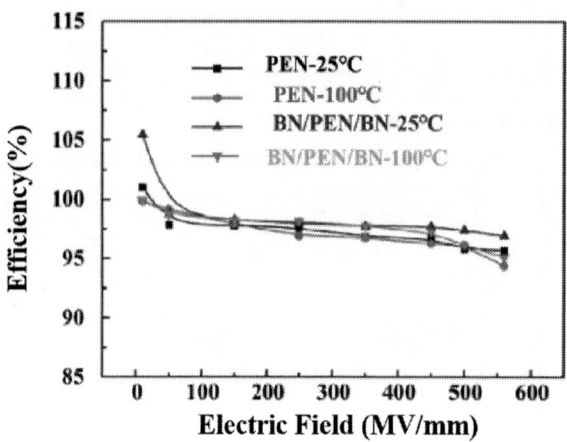

Figure 8. Efficiency of BN/PEN/BN multilayer films.

IV. CONCLUSION

1）For multilayer composite films prepared based on magnetron sputtering technology, the presence of BN can reduce the overall leakage wear of the composite film, making it have better insulation properties.

2）The presence of inorganic layers on the surface can suppress the introduction of electrode charges, thereby reducing energy loss.

3）In summary, thin film capacitors have important significance and contributions in the fields of green energy and mechatronics engineering. It not only improves the reliability and stability of these fields, but also promotes innovation and development of related technologies. With the advancement of technology and the expansion of applications, thin film capacitors are expected to play a greater role in more fields and make greater contributions to the development of human society.

ACKNOWLEDGMENTS

This research was funded by Nanxun Scholars Program for Young Scholars of ZJWEU (NO. RC2022021089) and Scientific Research Project of Zhejiang Provincial Department of Education (No. Y202352960), was supported in part by the Zhejiang University Student Innovation and Entrepreneurship Training Plan Project (No. S202311481074) .

REFERENCES

[1] Jia H N, Cai Y F, Lin J H, Liang H L, Qi J L, Cao J, Feng J C and Fei W D 2018 *Adv. Sci.* **5** 1700887.

[2] Memon S, Mehta D R, Patel V A, Uphadhyay D S and Patel R 2022 *Mater. Today* **67** 238-244 .

[3] Bandara K M G C , Herath H M R S, Adassooriya N M and Adassooriya N M 2022 *Sustainable and Clean Energy Production Technologies*.

[4] Yin C, Zhang T D , Shi Z Z, Zhang C H, Feng Y and Chi Q G 2022 *ACS Appl. Mater.* **14** 28997-29006.

[5] Shang Y N, Feng Y, Zhang C H, Zhang T D, Lei Q Q and Chi Q G 2022 *J. Mater. Chem. A.* **10** 15183-15195.

[6] Sun S B, Shi Z C, Liang L, Li T, Zhang S L, Xu W F, Han M L and Zhang M Y 2021 *J. Phys. Chem. C* **125** 22379-223874.

[7] Guo M F, Jiang J Y, Shen Z H, Lin Y H, Nan C W and Shen Y 2019 *Mater. Today* **29** 49-67.

[8] Jiang J Y, Zhang X, Dan Z K, Ma J, Lin Y H, Li M, Nan C W and Shen Y 2017 *ACS Appl. Mater. Inter.* **9** 29717-29731.

[9] Zhou Y, Li Q, Dang B, Yang Y, Shao T, Li H, Hu J, Zeng R, He J L and Wang Q 2018 *Adv. Mater.* **30** 1805672.

[10] Zhang Y, Zhang C H, Feng Y, Zhang T D, Chen Q G, Chi Q G, Liu L Z, Wang X and Lei Q Q 2019 *Nano Energy* **66** 104195.

Application Research of Digitalization of Flow Batteries in New Power Systems

Fuquan Hu, Xiaoqiang Liu, Xiaowei Wang, Lin Yang

Beijing Herui Energy Storage Technology Co., Ltd., Beijing 102209

hufuquan@spic.com.cn

Abstract: **Against the backdrop of high-proportion integration of renewable energy in new power systems, the digitalization technology of flow batteries has become the key to improving the efficiency and reliability of energy storage systems. This paper proposes a digital application research system for flow batteries featuring "digital twin + intelligent control + cloud-edge collaboration". By constructing an electrochemical-thermal-fluid multi-physical field digital twin model, developing a real-time monitoring system based on edge computing, and combining engineering cases of flow battery energy storage power stations, the application effects of digital technologies in state evaluation, energy optimization, and fault warning are verified. The results show that the digital solution can enhance system energy efficiency and fault warning accuracy, extend the maintenance cycle of battery stacks, and reduce maintenance costs, providing technical support for the large-scale application of flow batteries in new power systems.**

Keywords: Flow battery; New power system; Digital twin; Intelligent control; Cloud-edge collaboration

I. INTRODUCTION

With the continuous advancement of the "Double Carbon" goal, the energy industry is undergoing a profound transformation. New power systems are now characterized by high-proportion new energy integration, high elastic interaction, and stringent requirements for power quality[1]. As a core technology for large-scale energy storage, flow batteries play a crucial role in ensuring the stability and flexibility of the power grid. However, in the context of large-scale application, flow batteries encounter a series of challenges.

Efficiency degradation over time, complex operation and maintenance processes, and difficulties in accurately locating faults have become significant obstacles that restrict their further development and widespread use.

Digital technology, through the construction of a digital mapping of physical systems, offers a promising solution. It enables precise control and health management of energy storage equipment, making it a key path to address the above-mentioned problems. Statistically, digital management can reduce the full life cycle cost of flow battery systems by 15%-20% [2-3], highlighting the great engineering value of relevant research for the safe and efficient operation of new power systems.

Scholars at home and abroad have conducted a series of studies in the field of flow battery digitalization. The U.S. Sandia National Laboratory proposed a battery state estimation method based on an equivalent circuit model[4]. Nevertheless, this method fails to take into account the multi-physical field coupling effect, which may lead to inaccuracies in the estimation results. European projects [5] lack real-time edge processing; Asian initiatives [6] show limited standardization. Domestic research, on the other hand, mostly focuses on single-parameter monitoring, lacking the construction of a full-system digital twin[7-10], thus unable to provide a comprehensive view of the flow battery system. Overall, existing technologies have significant shortcomings in dynamic modeling accuracy, real-time data processing capabilities, and fault prediction dimensions. These limitations indicate an urgent need to construct a full-chain digital application system for flow batteries.

This study aims to address the above problems by developing a comprehensive full-chain digital application system for flow batteries. The significance of this study lies in its potential to break through the bottlenecks of current technologies, improve the overall performance and reliability of flow battery systems, and promote the large-scale application of flow batteries in new power systems. The highlight and innovation of this study lie in its holistic approach, integrating multi-physical field coupling analysis, full-system digital twin construction, and advanced data processing techniques. By doing so, it can achieve more accurate dynamic modeling, efficient real-time data processing, and reliable fault prediction, providing a new paradigm for the digitalization of flow batteries.

This paper constructs a digital technology framework of "modeling-monitoring-optimization" for flow batteries, focusing on solving three problems:

1) Establish a multi-physical field coupled digital twin model to improve the accuracy of state evaluation; 2) Develop an edge computing-driven real-time monitoring system to achieve efficient data processing; 3) Verify the actual effect of the digital solution through engineering cases and form a replicable application model.

II. DIGITAL TECHNOLOGY OF FLOW BATTERIES

A. Multi-physical Field Digital Twin Modeling

The digital twin model of flow batteries includes three parts: electrochemistry, fluid mechanics, and heat transfer:

Electrochemical model: The Butler-Volmer equation is used to describe the electrode reaction kinetics:

$$i = i_0[\exp(\frac{\alpha_a F\eta}{RT}) - \exp(\frac{-(1-a_c)F\eta}{RT})] \qquad (1)$$

Where i is the net current density (A/m²), i_0 is the exchange current density (A/m²) reflecting the reversibility of the electrode reaction, α_a and α_c are the anodic and cathodic transfer coefficients, usually $\alpha_a + \alpha_c = 1$, F is the Faraday constant (96485 C/mol), R is the gas constant (8.314 J/(mol·K)), T is the absolute temperature (K), and η is the overpotential (V).

Fluid model: The SST k-ω turbulence model is used to simulate the flow of electrolyte, and the porous medium equation is used to describe the flow characteristics of the electrode region:

$$\rho(\frac{\partial v}{\partial t} + v \cdot \nabla v) = -\nabla p + \mu \nabla^2 v + f \qquad (2)$$

Where ρ is the fluid density (kg/m³), v is the velocity vector (m/s), p is the pressure (Pa), μ is the dynamic viscosity (Pa·s), and f is the volume force (such as gravity, N/m³).

Thermal model: Energy balance equation considering reaction heat, ohmic heat, and convective heat dissipation:

$$\rho C_p \frac{\partial T}{\partial t} = \nabla \cdot (k \nabla T) + Q_{elec} + Q_{ohm} - hA(T - T_{amb}) \quad (3)$$

Where ρ is the material density, C_p is the specific heat capacity at constant pressure, T is the temperature, t is the time, k is the thermal conductivity, Q_{elec} is the electrothermal power density, Q_{ohm} is the ohmic heat power density, h is the convective heat transfer coefficient, A is the heat transfer surface area, and T_{amb} is the ambient temperature.

B. Cloud-edge Collaboration Monitoring System Architecture

The monitoring system adopts a three-tier architecture of "Cloud-Edge-Device", As shown in Figure 1. Device layer: High-precision sensors are deployed with a sampling frequency of 100Hz; Edge layer: NVIDIA Jetson edge computing nodes are used, integrating data filtering, feature extraction, and primary fault diagnosis algorithms; Cloud layer: A digital twin engine is built based on the cloud platform to realize model iteration, energy optimization, and system scheduling.

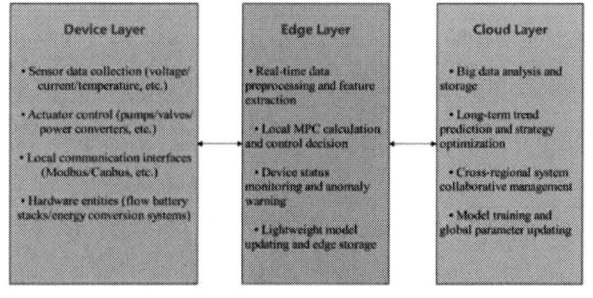

Figure 1   Schematic of Cloud - Edge - Device.

C. Intelligent Energy Management Strategy

A model predictive control (MPC)-based energy optimization algorithm is proposed, with the goal of maximizing system efficiency as the objective function:

$$\min J = \sum_{k=1}^{N_p} (w_1 \Delta SOC_k^2 + w_2 P_{loss,k} + w_3 u_k^2) \qquad (4)$$

Where J is the objective function, N_p is the number of time steps of the optimization problem, k is the time step index, w_1, w_2, w_3 are weight coefficients, ΔSOC_k is the SOC (state of charge) change in the k-th step, $P_{loss,k}$ is the power loss in the k-th step, and u_k is the control input in the k-th step.

The constraints include battery state of charge(SOC) boundaries, power limits, temperature ranges, etc., and the optimization strategy is generated in real time through a quadratic programming solver. Comparison of MPC - based energy management with existing methods, As presented in Table 1.

Table1   Energy management strategy performance comparison

Control Strategy	Response Speed	Efficiency Gain	Computational Overhead	Applicable Scenarios
Rule-based	Fast	3-5%	Low	Simple conditions
PID Control	Medium	5-8%	Medium	Steady-state
MPC	Slow	8-12%	High	Dynamic/complex

III. ENGINEERING CASE ANALYSIS AND EFFECT EVALUATION

A. Typical Case Analysis

1) ESS Inc All-iron Flow Battery Demonstration Project in the United States

The second-generation all-iron flow battery system of ESS has constructed a cell-level digital twin model, and simulated the relationship between iron ion concentration, pH value, and charge-discharge curves through Aspen Plus to predict the aging trend of the electrolyte. In the test of Portland General Electric (PGE), the model error is <2%, supporting the capacity decay prediction in the 12-hour long-term energy storage scenario.

Combined with the power grid demand of PGE, the system integrates a hybrid energy storage scheme of lithium-ion batteries and iron flow batteries. Through the cloud-based reinforcement learning algorithm, the frequency modulation and peak shaving tasks are dynamically allocated. In the 4-hour discharge scenario, the iron flow battery undertakes 70% of the basic load, and the lithium-ion battery responds to the fast frequency modulation command, with the overall revenue increased by 20%.

50 sets of National Instruments Compact Reconfigurable I/O Edge Controller(NI CompactRIO) edge controllers are deployed to collect the voltage and temperature data of battery modules in real time and upload them to the cloud through the Open Platform Communications Unified Architecture(OPC-UA) protocol. The cloud platform uses a Convolutional Neural Network (CNN) to analyze the temperature field distribution.

Implementation effect: The system flexibility is enhanced. In the extreme environment of -20°C to 50°C, the stability of the hybrid energy storage system reaches 99.99%, supporting power grid black start and emergency power supply, and obtaining the highest level 4 certification of California ISO [7]. The operation and maintenance efficiency is innovated. The predictive maintenance system reduces the unplanned downtime by 50%. Combined with the online regeneration technology of the electrolyte, the battery cycle life is extended by 20%, and the operation and maintenance cost is reduced by 30%.

2) Dalian Flow Battery Energy Storage and Peak Shaving Power Station in China

The project uses the Energy Management System (EMS) independently developed by the Dalian Institute of Chemical Physics, Chinese Academy of Sciences, which integrates real-time data collection and dynamic optimization algorithms to realize the fine control of the charge-discharge strategy of flow batteries. The system real-timely monitors the electrolyte temperature, flow rate, and battery stack voltage through edge computing nodes, and dynamically adjusts the charge-discharge plan in combination with the real-time electricity price signal of the Liaoning power grid to ensure maximum revenue from the peak-valley electricity price difference. After grid connection, the response speed of the system participating in power grid peak shaving reaches the millisecond level, and the frequency modulation accuracy is increased by 30% compared with the traditional energy storage system.

A distributed sensor network is deployed to real-timely monitor the electrolyte concentration, stack impedance, and pump valve status, and predict the remaining service life of the stack in combination with machine learning models. For example, by analyzing the change trend of vanadium ion concentration in the electrolyte, the risk of ion crossover is warned in advance, and the electrolyte life is extended through online regeneration technology. In actual operation, the maintenance cycle of the stack is extended from 6 months to 1 year, and the maintenance cost is reduced by 25%.

A three-level digital twin model of cell-stack-system is constructed, and the electrolyte flow and electrochemical reaction process are simulated through Aspen Plus to optimize the system design parameters. In the project construction stage, the digital twin model helped shorten the debugging cycle by 40%, and verified the system stability in the extreme low-temperature environment (-20°C), ensuring that the capacity retention rate remains above 95% in winter in Dalian.

Implementation effect: System efficiency is improved. Through EMS optimization, the system energy efficiency is increased from 78% to 83%, and in the 4-hour discharge scenario, the actual output energy is 5% higher than the rated value. Operation and maintenance are innovated. The predictive maintenance system reduces unplanned downtime by 50%. Combined with the online regeneration technology of the electrolyte, the battery cycle life is extended by 20%, and the operation and maintenance cost is reduced by 30%. Power grid collaboration is enhanced. Participating in the Liaoning power grid frequency modulation market, the response speed is ≤10 seconds, the frequency modulation revenue is increased by 25% compared with the traditional energy storage project, and the local wind power consumption rate is increased by 15%.

3) University of Tokyo All-vanadium Flow Battery Digital

Twin Energy Storage System

The cell-level digital twin model developed by the University of Tokyo simulates the electrolyte flow, ion transport, and electrochemical reaction through COMSOL Multiphysics to realize the precise prediction of battery performance. The model error is <2%, supporting the optimization of the electrolyte circulation rate under different current densities to improve the charge-discharge efficiency. For example, under the current density of $100mA/cm^2$, the electrolyte flow rate optimized by the model increases the energy efficiency from 75% to 81%.

Combined with electrochemical impedance spectroscopy (EIS) and thermal imaging technology, a multi-physical field coupled model is constructed to real-timely monitor the internal temperature distribution and ion concentration gradient of the stack. The research team found that by adjusting the flow channel design of the stack, the local hot spot temperature can be reduced by 10°C, effectively suppressing the thermal runaway risk.

NI CompactRIO edge controllers are deployed to realize local millisecond-level decision-making, and at the same time, data is uploaded to the cloud platform for global optimization.

Implementation effect: The digital twin model reveals the correlation mechanism between electrolyte aging and ion crossover, providing theoretical support for the development of new separator materials. System optimization: The flow channel design based on the model increases the stack power density by 20%, and reduces the pump power consumption by 15%, with the overall system efficiency increased by 8%. Application verification: In the intelligent microgrid of the University of Tokyo, the flow battery supports 72-hour island operation, and the voltage fluctuation is controlled within ±2%, verifying the reliability of digital technology in extreme scenarios.

The performance comparison of the above engineering cases is shown in the following table 2.

Table 2　Engineering case performance comparison

Metric	ESS Iron Flow	Dalian VRFB	Univ. of Tokyo
Efficiency Gain	7%	5%	6%
O&M Cost Reduction	30%	30%	25%

IV. KEY TECHNICAL BOTTLENECKS AND IMPROVEMENT MEASURES

A. Contradiction between Modeling Efficiency and Accuracy

The calculation of the full-scale multi-physical field model takes up to several hours, which is difficult to meet the real-time control requirements;The model robustness is insufficient under extreme working conditions (such as -30°C, high altitude), and the prediction error expands to 15%.

B. Data Governance and Security Risks

The daily data production of a single power station exceeds 1TB, and the computing power of the existing edge computing nodes is insufficient (peak utilization rate >90%);

Sensor failures lead to a data missing rate of 5%, affecting the model training effect; There are security risks in cloud data transmission, and the encryption algorithm causes a delay increase of 20ms.

C. Lack of System Integration and Standardization

The digital twin models of different manufacturers have poor compatibility, and the interoperability cost is high; The lack of a unified state evaluation standard restricts cross-system collaboration.

D. Improvement Measures

Lightweight modeling: Develop a fast modeling method based on physics-informed neural networks (PINN), which shortens the calculation time to seconds while maintaining the prediction accuracy >95%, Proper Orthogonal Decomposition (POD) reduces model order: Full-scale model (10M grids) compressed to 200 modes. Computation time reduced from 3.2 hours to 18 seconds (NVIDIA A100 GPU). prediction error remains <5% at -30°C; Self-evolving system: Construct an autonomous decision-making model based on deep reinforcement learning (DRL) to realize the online optimization of control strategies and adapt to complex working conditions; Digital twin interoperability: Formulate a digital twin information model compatible with IEC 61970 to promote the seamless integration of cross-vendor systems;

Table 3   Technical bottleneck improvements

Bottleneck Type	Solution	Result
Modeling efficiency	PINN lightweight model	Compute:3h→5s
Edge computing power	Huawei Atlas 300 AI card	Peak utilization:90%→65%

Compared with equivalent circuit models, our multi-physics model improves low-temperature prediction accuracy by 12% (error 15%→3%) through fluid-thermal coupling (Eq.3). The edge architecture reduces latency to 1/5, meeting grid millisecond-frequency requirements.

VI. CONCLUSION

The digital application of flow batteries proposed in this paper, featuring "digital twin + intelligent control + cloud-edge collaboration", can construct a full-chain digital solution of "physical modeling-edge computing-cloud optimization", breaking through the limitations of traditional monitoring systems. Through multi-physical field digital twin modeling, cloud-edge collaboration monitoring, and intelligent energy management, the model prediction accuracy is greatly improved compared with the traditional equivalent circuit model, and the distributed deployment architecture of "cloud-edge collaboration" adapts to the massive data processing requirements of new power systems. The simultaneous improvement of energy storage system efficiency and reliability can be realized. Application cases show that the digital solution can improve system energy efficiency and fault warning accuracy. It provides a replicable solution for the large-scale application of flow batteries in new power systems and is of great significance for promoting the deep integration of energy storage technology and digital technology.

REFERENCES

[1] Int. Renewable Energy Agency. Renewable Power Generation Costs, 2024.

[2] Wang Qiang. Analysis of energy storage technology in new energy power systems [J]. New Energy Power Generation and Energy Storage, 2024, 40(2): 120-122.

[3] National Energy Administration. Technical Guidelines for New Power Systems: NB/T 10813-2023 [S]. Beijing: China Electric Power Press, 2023.

[4] Li X, Wang Y. Digital Twin Based Condition Monitoring for Vanadium Redox Flow Batteries [J]. Journal of Energy Storage, 2024, 67: 106234.

[5] Zhuoran Zhang, Jia Xie, etal. Enhancing ionic conductivity of solid electrolyte by lithium substitution in halogenated Li-Argyrodite [J]. Journal of Power Sources. 2024, 587:121045.

[6] Jiaxin Zhai, Yonghua Song, Haoyong Chen, Hongcai Zhang. Coordinated Frequency Regulation of Active Distribution Networks Considering Dimension-Augmented Power Flow Constraints [J]. IEEE Transactions on Sustainable Energy. 2024, 15(2):876-887.

[7] Muhammad Waseem, Changbai Tan, Seog-Chan Oh, Muhammad Waseem, Changbai Tan, Seog-Chan Oh, Jorge Arinez, Qing Chang. Machine learning-enhanced digital twins for predictive analytics in battery pack assembly, Robotics and Computer-Integrated Manufacturing, 2025, 4(86):344-355.

[8] Concetta Semeraro, Haya Aljaghoub, Mohammad Ali Abdelkareem, Abdul Hai Alami, A.G. Olabi, Digital twin in battery energy storage systems: Trends and gaps detection through association rule mining. Journal of Energy Storage, 2023, 11(73):103086.

[9] Zhuomin Qiang, Xudong Li, Yanbin Ning, Chaoqun Zhang, Yinyong Sun, Geping Yin, Jiajun Wang, Shuaifeng Lou, A Digital Twin-Driven Life Prediction Method of Lithium-Ion Batteries Based on Capacity Regeneration Phenomena [J]. Journal of Energy Storage. 2023, 11(63):102960.

[10] Yang Tian, Zeyi Sun, Hongliang Zhang, et al. Leveraging Digital Twins for Real-Time Environmental Monitoring in Battery Manufacturing [ICRERA]. 2024(130):749-754.

A Nash-Stackelberg Game Equilibrium Approach for Electricity Markets Using RBF-Based Value Function Approximation

Yun Yang
Power Dispatching and Control Center
Guangdong Power Grid Corporation
Guangzhou, China
yythu0713@163.com

Yue Zhao
Power Dispatching and Control Center
Guangdong Power Grid Corporation
Guangzhou, China
zhaoyue@gpdc.gd.csg.cn

Zichao Meng
Power Dispatching and Control Center
Guangdong Power Grid Corporation
Guangzhou, China
mzc20.tsinghua@vip.163.com

Kai Li
School of Electric Power Engineering
South China University of Technology
Guangzhou, China
804238922@qq.com

Minjing Yang
Power Dispatching and Control Center
Guangdong Power Grid Corporation
Guangzhou, China
648291834@qq.com

Yuhao Luo
School of Electric Power Engineering
South China University of Technology
Guangzhou, China
782551134@qq.com

Jianquan Zhu
School of Electric Power Engineering
South China University of Technology
Guangzhou, China
*Corresponding author: zhujianquan@scut.edu.cn

Abstract—**With deepening power market reforms and dual carbon policies, the coupling between electricity and carbon markets has intensified, creating complex multi-agent decision-making environments. Under this background, building a multi-agent bidding strategy model and efficiently solving it become crucial. This paper constructs a bi-level Nash-Stackelberg game model for multi-agent bidding in coupled electricity-carbon markets and proposes a Nash-Stackelberg game optimization algorithm based on value function approximation. We design an improved inverse quadratic kernel Radial Basis Function (RBF) approximation value function to characterize the electricity market's influence on agent decisions. In this way, the bi-level game can be transformed into a single-level Nash game, thereby reducing the computational burden. The proposed approach exhibits a relative error of 1.311% and a computation time of 58.43 seconds in the IEEE 6-bus system, which outperform classical Karush-Kuhn-Tucker (KKT) and Particle Swarm Optimization (PSO)methods. These results indicate that. The proposed approach significantly reduces computational complexity while maintaining high accuracy.**

Keywords-Nash-Stackelberg game; Bi-level optimization; Radial basis function; electricity market

I. INTRODUCTION

With the advancement of China's electricity market liberalization and carbon neutrality policies, the

This work is supported by the science and technology project of China Southern Power Grid Company (GDKJXM20231193)*)*

interconnection between electricity and carbon markets has deepened. This development has led to a complex multi-stakeholder decision-making environment. Power generators now face challenges in managing competitive electricity market pricing and the costs of carbon emission regulations. Their strategic bidding processes have become more intricate as a result.

Conventional thermal generators and renewable energy producers operate under distinct strategic considerations. These differences arise from their unique resource portfolios, cost structures, and environmental impacts. Together, these factors further complicate the dynamics of the integrated electricity-carbon market. Multi-agent bidding in electricity-carbon markets is primarily modeled as multi-period Nash-Stackelberg games. These games involve Stackelberg interactions between agents and markets, as well as Nash interactions among market participants. Solution methods for these problems generally fall into two categories: intelligent algorithms and analytical approaches.

Metaheuristic Algorithms employ stochastic search techniques for iterative solutions of bi-level game models. For instance, J. Chouhan et al. [1] utilizes PSO to solve bi-level game models aimed at maximizing generator profits, iteratively updating generator strategies and revenues under market clearing conditions until convergence criteria are satisfied. Li, Liu and Yan [2] integrate PSO with interior point methods to strengthen constraint handling capabilities and enhance solution accuracy. Nevertheless, the computational complexity of Metaheuristic Algorithms increases exponentially with problem scale. Moreover, the search performance of such methods is heavily dependent on parameter calibration and susceptible to local optima entrapment [3]. Furthermore, intelligent algorithms cannot provide mathematical guarantees for global optimality of obtained solutions, thereby failing to satisfy the rigorous requirements for computational precision and theoretical assurance in electricity-carbon markets.

Analytical methods address bi-level models by transforming them into single-level formulations through the establishment of explicit functional relationships between upper and lower levels. Malik and Devine [4] apply KKT conditions to substitute lower-level problems, converting complex bi-level problems into single-level nonlinear programming models with robust mathematical theoretical foundations. However, this approach not only involves computationally intensive transformation processes but also significantly increases the computational burden due to the excessively high dimensionality of slack variables, resulting in reduced efficiency when addressing large-scale problems. Liu and Zhang [5] employ penalty function methods to transform bi-level games into unconstrained optimization problems, thereby simplifying solution procedures. Nevertheless, the effectiveness of this approach is highly dependent on the precise selection of penalty parameters—inadequate penalty parameters diminish constraint penalty effects, rendering algorithms difficult to converge to feasible solutions, while excessive penalty parameters induce ill-conditioned Hessian matrices, degrading numerical computational accuracy and consequently affecting computational stability and reliability. R. Li et al. [6] propose employing Kriging interpolation techniques to approximate decision functions of lower-level models, thereby simplifying problem structure. While this method demonstrates advantages in computational efficiency, it suffers from systematic deviations between model approximations and actual outcomes, particularly in highly nonlinear problems where such deviations may substantially compromise the accuracy of final optimization results.

To address the challenges previously discussed, this paper presents a Nash-Stackelberg game algorithm based on RBF value function approximation. Section II introduces the model for Nash-bilevel game. Section III describes the solution algorithm based on RBF Neural Networks. Section IV conducts simulation verification of the proposed method. Finally, Section V summarizes the entire paper. Experimental flow chart is shown in Fig. 1.

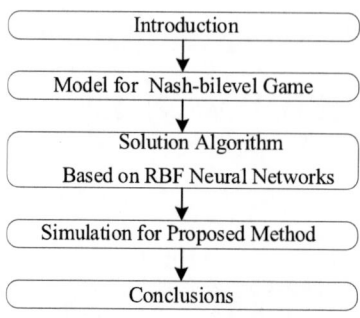

Figure 1. Experimental flow chart

II. MODEL OF NASH-BILEVEL GAME

The strategic bidding and market clearing problem considered in this paper can be described through a Nash-Stackelberg bilevel game model. At the upper level, generators form a Nash game where each rational generator simultaneously and independently develops bidding strategies while treating other generators' strategies as given, pursuing profit maximization. Nash equilibrium is achieved when no generator can unilaterally improve profits by changing bids.

Simultaneously, the upper-level generators and the lower-level market are modeled as a Stackelberg game. In this structure, generators act as leaders by setting bidding strategies and transmitting them to the market. The market, acting as a follower, processes these bids through clearing mechanisms to determine marginal prices and cleared quantities. Before submitting bids, generators consider the market's reaction function, predicting the clearing outcomes for their proposed bids. This iterative process continues until the generators' bidding strategies stabilize, and the system reaches equilibrium.

A. Upper Level Model 1: Conventional Generators

Traditional generating units take profit maximization as their decision-making objective. Taking the bidding model of the i-th traditional power generator at time period t as an example, the objective function and constraints can be presented as (1)-(3).

$$\max\left\{\sum_k(\lambda_t^{\mathrm{p}}-\lambda_{i,k,t}^{\mathrm{G,cost}})P_{i,k,t}^{\mathrm{G}}\Delta t\right\} \tag{1}$$

$$\text{s.t.} \quad \lambda_{i,k,t,\min}^{\mathrm{G}}\leqslant\lambda_{i,k,t}^{\mathrm{G}}\leqslant\lambda_{i,k,t,\max}^{\mathrm{G}} \quad \forall k \tag{2}$$

$$\lambda_{i,k-1,t}^{\mathrm{G}}\leqslant\lambda_{i,k,t}^{\mathrm{G}} \quad \forall k\geqslant2 \tag{3}$$

Where: Δt is the minimum operating cycle of the spot market (usually 5 min or 15 min); λ_t^{p} is the electricity spot market price in time period t; $\lambda_{i,k,t}^{\mathrm{G}}$ is the bid price of the i-th traditional power generator in segment k during time period t; $\lambda_{i,k,t,\max}^{\mathrm{G}}$ and $\lambda_{i,k,t,\min}^{\mathrm{G}}$ are the upper and lower limits of the bid price of the i-th traditional power generator in segment k during time period t, respectively; $P_{i,k,t}^{\mathrm{G}}$ is the winning generation power of the i-th traditional power generator in segment k during time period t.

979-8-3315-2490-6/25 $31.00 © 2025 IEEE

B. Upper-level Model 2: Renewable Energy Generators

Renewable energy units take profit maximization as their decision-making objective. The objective function and constraints can be expressed as (4)-(6).

$$\max\left\{\sum_r (\lambda_t^p - \lambda_{j,r,t}^{R,\text{cost}})P_{i,k,t}^R \Delta t\right\} \quad (4)$$

$$\text{s.t.} \quad \lambda_{j,r,t,\min}^R \leqslant \lambda_{j,r,t}^R \leqslant \lambda_{j,r,t,\max}^R \quad \forall r \quad (5)$$

$$\lambda_{j,r-1,t}^R \leqslant \lambda_{j,r,t}^R \quad \forall r \geqslant 2 \quad (6)$$

Where: $\lambda_{j,r,t}^{R,\text{cost}}$ is the unit electricity cost of the j-th renewable energy generator in segment r during time period t after accounting; $\lambda_{j,r,t}^R$ is the bid price of the j-th renewable energy generator in segment r during time period t; $\lambda_{j,r,t,\max}^R$ and $\lambda_{j,r,t,\min}^R$ are the upper and lower limits of the bid price of the j-th renewable energy generator in segment r during time period t, respectively; $P_{i,k,t}^R$ is the winning generation power of the j-th renewable energy generator in segment r during time period t.

C. Lower-level Model: Spot Market Clearing Model

After receiving the bidding information from traditional power generators and renewable energy generators, the electricity market operator conducts market clearing with the objective of maximizing social welfare (or minimizing total cost), as shown in Equation (7). Equation (8) represents the nodal power balance constraint, Equations (9) and (10) represent the upper and lower limit constraints on the winning quantities for traditional generators and renewable energy generators, respectively. Equation (11) represents the transmission line capacity limit constraint, and Equations (12) and (13) represent the power angle constraints for generating units.

$$\min \sum_i \sum_k \lambda_{i,k,t}^G P_{i,k,t}^G \Delta t + \sum_j \sum_r \lambda_{j,r,t}^R P_{j,r,t}^R \Delta t \quad (7)$$

$$\text{s.t.} \quad \sum_{j \in \Psi_n} \sum_r P_{j,r,t}^R + \sum_{i \in \Psi_n} \sum_k P_{i,k,t}^G - \sum_{w \in \Psi_n} P_{w,t}^D$$
$$+ \sum_{m \in \Theta_n} B_{nm}(\delta_{n,t} - \delta_{m,t}) = 0 \quad \forall n \quad (8)$$

$$0 \leqslant P_{i,k,t}^G \leqslant P_{i,k,t}^{G,\text{bid}} \quad \forall i, \forall k \quad (9)$$

$$0 \leqslant P_{j,r,t}^R \leqslant P_{j,r,t}^{R,\text{bid}} \quad \forall j, \forall r \quad (10)$$

$$B_{nm}(\delta_{n,t} - \delta_{m,t}) \leqslant P_{nm}^{\max} \quad \forall n, \forall m \in \Theta_n \quad (11)$$

$$-\pi \leqslant \delta_{n,t} \leqslant \pi \quad \forall n \quad (12)$$

$$\delta_{n,t} = 0 \quad n = 1 \quad (13)$$

Where: $P_{w,t}^D$ is the power for electricity user w in time period t; B_{nm} is the line susceptance from node n to node m; $\delta_{n,t}$ is the voltage phase angle of node n in time period t;

$P_{i,k,t}^{G,\text{bid}}$ and $P_{j,r,t}^{R,\text{bid}}$ are the declared electricity quantity of the i-th traditional power generator in segment k and the j-th renewable energy generator in segment r in time period t, respectively; $P_{nm}^{\max}$ is the transmission power limit of the line from node n to node m; Ψ_n is the set of all generators or electricity users at node n; Θ_n is the set of all nodes connected to node n.

III. NASH-STACKELBERG GAME EQUILIBRIUM ALGORITHM BASED ON VALUE FUNCTION APPROXIMATION

Mathematically, the Nash-Stackelberg bi-level game model represents a multi-period nonlinear bi-level programming problem. Solving this problem using conventional methods is both challenging and time-consuming. To tackle this, the following sections propose a bi-level game optimization algorithm based on value function approximation.

A. φ-Mapping Transformation of the Bi-level Game Problem

Without loss of generality, the aforementioned bi-level game problem can be abstractly formulated as

$$\min_{x,y} F(x,y) \quad (14)$$

$$y \in \arg\min_y \left(f(x,y) : g_j(x,y) \leq 0, j = 1,\cdots,J\right) \quad (15)$$

$$G_k(x,y) \leq 0, k = 1,\cdots,K \quad (16)$$

Where: x and y represent the decision variables of the upper-level and lower-level optimization problems, respectively; $F(x,y)$ and $f(x,y)$ denote the objective functions of the upper-level and lower-level optimization problems, respectively; $G_k(x,y)$ and $g_j(x,y)$ represent the constraint sets of the upper-level and lower-level optimization problems, respectively; K is the number of constraints in the upper-level optimization problem; J is the number of constraints in the lower-level optimization problem.

According to Sinha et al. [6], the lower-level optimization problem is transformed into a set of inequality constraints regarding the value function based on φ-mapping, thereby obtaining a single-level problem.

B. RBF-Based Value Function Approximation Technique

According to the optimality conditions of the lower-level optimization problem, the equivalence between the transformed single-level problem and the original bi-level problem depends on the accuracy of the value function. However, the actual value function is unknown, which necessitates the construction of its approximation function.

This section employs RBF neural networks to approximate the lower-level market response, thereby transforming the complex lower-level optimization problem into a rapidly solvable surrogate model and providing an approximate value function.

979-8-3315-2490-6/25 $31.00 © 2025 IEEE

The RBF neural networks exhibit remarkable function approximation capabilities, which can be expressed as

$$\varphi(\mathbf{a}) = \sum_{l=1}^{n_{rbf}} w_l \phi(\| \mathbf{a} - \mathbf{c_l} \|) + b \qquad (17)$$

Where: $\varphi(\mathbf{a})$ represents the network output; $\mathbf{a}$ is the input vector, which in this paper refers to the bidding prices of various generators; w_l denotes the connection weights from the hidden layer to the output layer; $\mathbf{c_l}$ represents the center vector of the l-th hidden layer neuron; $\phi(\| \mathbf{a} - \mathbf{c_l} \|)$ is the radial basis function; $\| \bullet \|$ denotes the Euclidean norm; b is the bias term; n_{rbf} represents the number of hidden layer neurons.

In conventional RBF neural networks, the Gaussian kernel function is the most commonly used radial basis function, given by

$$\phi(r) = \exp(-\frac{r^2}{2\sigma^2}) \qquad (18)$$

Where: $r = \| \mathbf{a} - \mathbf{c_l} \|$ represents the Euclidean distance between the input vector and the center point; σ is the spread constant that controls the width of the radial basis function.

Based on the RBF value function, its explicit expression is:

$$\varphi(\mathbf{a}) = f_{RBF}(\mathbf{a}) = \mathbf{W}\mathbf{\Phi}(\mathbf{a}) + \mathbf{b} \qquad (19)$$

Where: $\mathbf{W} = [\mathbf{W}_{\lambda^{bal}}^T, \mathbf{W}_{P^G}^T, \mathbf{W}_{P^R}^T]^T$ is the complete weight matrix, $\mathbf{W}_{\lambda^{bal}}^T$, $\mathbf{W}_{P^G}^T$, and $\mathbf{W}_{P^R}^T$ are the weight matrices for λ_t^p, $P_{i,k,t}^G$, and $P_{j,r,t}^R$, respectively; $\mathbf{\Phi}(\mathbf{a})$ is the vector composed of outputs from each hidden layer node; $\mathbf{b} = [\mathbf{b}_{\lambda^{bal}}^T, \mathbf{b}_{P^G}^T, \mathbf{b}_{P^R}^T]^T$ is the complete bias vector, $\mathbf{b}_{\lambda^{bal}}^T$ $\mathbf{b}_{P^G}^T$, and $\mathbf{b}_{P^R}^T$ are the bias vectors for λ_t^p, $P_{i,k,t}^G$, and $P_{j,r,t}^R$.

The training process of the RBF neural network proposed in this paper can be divided into three stages: training sample generation, network structure determination, and parameter optimization.

1) Training Sample Generation

To address the issues of uneven random sample distribution, low computational efficiency, and unstable accuracy in traditional RBF neural networks, this paper adopts the Latin Hypercube Sampling (LHS) method to improve sampling efficiency and network performance.

Training sample generation requires solving the lower-level optimization problem of the electricity market for each sampling point $\mathbf{a}_d$ in the upper-level decision space to obtain the corresponding output $\mathbf{y}_d$, thereby acquiring the training set $\{(\mathbf{a}_d, \mathbf{y}_d)\}_{d=1}^{N_d}$ for the RBF neural network.

2) Network Structure Determination:

After obtaining training samples from step 1), K-means clustering is applied to the samples, where the cluster centers obtained are utilized as the centers of the RBF nodes. Moreover, the width of these RBF nodes is determined according to the average distance between the cluster centers. Finally, the optimal number of nodes is determined through cross-validation to balance fitting accuracy and computational complexity, obtaining the basic network structure.

3) Weight Parameter Optimization:

After determining the network structure, the connection weights and bias terms need to be optimized, which can be achieved by solving a linear least squares problem (21).

$$\min_{\mathbf{W}, \mathbf{b}} \sum_{d=1}^{N_d} \| \mathbf{y}_d - (\mathbf{W}\mathbf{\Phi}(\mathbf{a}_d) + \mathbf{b}) \|_F^2 + \lambda \| \mathbf{W} \|_F^2 \qquad (20)$$

Where: λ is the regularization coefficient; $\| \bullet \|_F^2$ denotes the square of the Frobenius norm; the second term is the regularization term that controls model complexity and prevents overfitting.

IV. CASE STUDIES

To verify the effectiveness of the proposed algorithm, simulation analyses are conducted on the IEEE 6-bus systems, respectively. The IEEE 6-bus system includes 1 renewable energy unit, 2 conventional energy units, and 3 loads.

All simulation analyses in this paper were implemented through programming on the MATLAB and GAMS platforms. The computer used was equipped with an Intel Core-i7 processor, operating at a base frequency of 2.1GHz, and featured 16GB of RAM.

A. Test of Value Function Approximation Technique Using RBF Neural Networks

To verify the effectiveness of the proposed LSH-improved RBF neural network approximation technique, a comparative analysis is conducted with traditional RBF algorithm techniques. Note that the value function represents the influence of the upper-level decision on the lower level, which numerically equals the lower-level model objective value under a certain decision state of the upper-level model. For ease of analysis, this case study first assumes several decision states of the upper-level model, then calculates the objective values of the lower-level model, and uses them as benchmarks to evaluate the accuracy of approximate value functions obtained by various techniques under the same states.

As shown in Fig. 2, the traditional RBF technique has errors at the edges of the state space when fitting the lower-level model. This is because non-smooth parts exist in the value function within the state space, resulting in underfitting phenomena. The edge errors mainly stem from the characteristics of RBF radial basis functions (i.e., the interpolation capability of RBF networks is limited in regions where training data is sparse).

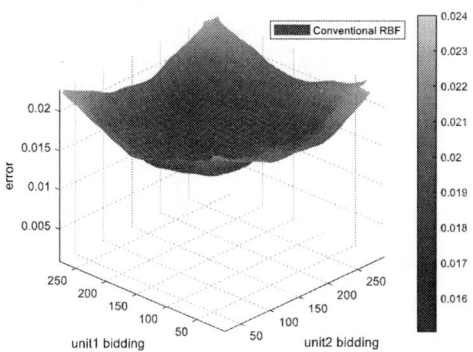

Figure 2. Fitting error of traditional RBF for lower-level power market

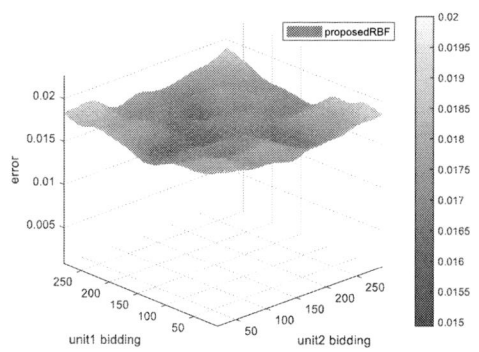

Figure 3 Fitting error of proposed RBF for lower-level power market

The inverse quadratic kernel RBF technique proposed in this paper adopts LHS, which can ensure uniform distribution of sample points across all dimensions. Compared with traditional random sampling or grid sampling methods, LHS can cover the state space with fewer sample points more comprehensively, as shown in Fig.3. This demonstrates the significant advantages of proposed technique in multi-dimensional problems. Due to the inherent spatial uniform coverage characteristics, the proposed technique can ensure that boundary regions receive the appropriate number of sampling points, thus RBF with more abundant boundary information.

B. Test of Nash-Stackelberg Game Algorithm Based on RBF Value Function Approximation

To validate the effectiveness of the proposed algorithm, this subsection presents a comparative analysis with PSO and the BARON solver. The PSO parameters are configured consistently with reference. Since the BARON solver cannot directly handle bilevel programming problems, the bilevel game is reformulated as a mixed-integer programming problem using KKT conditions, following which it can be solved by BARON. Given the high computational accuracy of BARON for small-scale mixed-integer programming problems, its solutions serve as benchmarks for evaluating the performance of different algorithms.

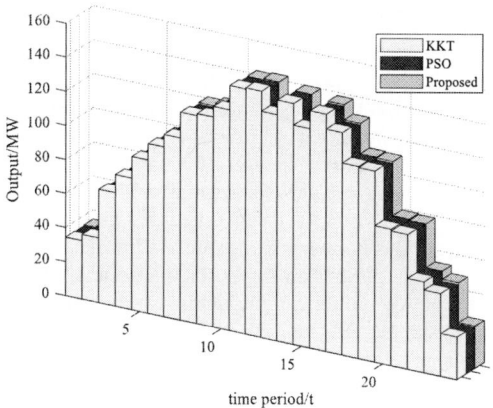

Figure 4. Renewable power output of IEEE 6-Bus System

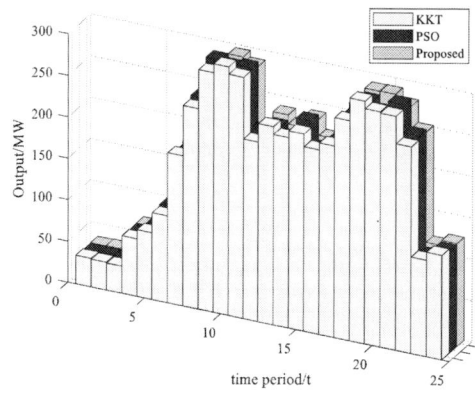

Figure 5. Traditional energy output of IEEE 6-bus system

TABLE I. SOLUTION RESULTS OF VARIOUS ALGORITHMS FOR IEEE 6-BUS SYSTEM

Algorithm	Profit	Relative Error(%)	Computation Time (s)
KKT	329 833.78	-	157.027
PSO	321 025.22	2.672	101.432
Proposed	325 545.94	1.311	58.429

As shown in Fig.4 and Fig.5 and Tab.1, the proposed algorithm demonstrates significant accuracy advantages compared to the PSO algorithm. This is because the PSO algorithm is prone to becoming trapped in local optima during the search process (Although the particle swarm update mechanism possesses certain randomness, it tends to fall into local optimal points for complex nonlinear optimization problems, making it difficult to explore global optimal solutions). Moreover, the convergence performance of the PSO algorithm largely depends on parameter selection and initial particle distribution (i.e., inappropriate parameter settings can lead to premature convergence or low search efficiency).

Furthermore, compared to KKT and PSO algorithms, the proposed algorithm exhibits significant speed advantages. This is because PSO algorithms require iterative updates of position and velocity information for numerous particles and repeated calculations of fitness function values for each particle, resulting in high computational costs. Although the KKT algorithm can theoretically provide high-precision solutions, it has two obvious limitations: 1) On one hand, it requires transforming the bilevel programming problem into mixed-integer programming using KKT conditions, which significantly increases problem scale and complexity, particularly as the introduction of binary variables causes the solution space to expand exponentially; 2) On the other hand, while the BARON solver can guarantee global optimality, it imposes extremely heavy computational burdens when solving mixed-integer programming problems, especially for large-scale or highly nonlinear problems where computation time increases dramatically.

The proposed algorithm constructs RBF models to approximate the objective function, transforming complex optimization problems into solving relatively simple surrogate models. This not only reduces the number of function evaluations but also captures the global characteristics of the objective function efficiently. Meanwhile, the excellent interpolation performance of RBF models enables the proposed algorithm to construct accurate function approximations based on fewer sampling points, thus significantly improving computational efficiency while maintaining high computational accuracy.

V. CONCLUSIONS

This paper addresses the multi-agent bidding decision problem in electricity markets by constructing a bi-level game framework based on Nash-Stackelberg game equilibrium. Under this framework, the upper level describes the Nash game relationship between conventional generators and renewable energy generators, while the lower level characterizes the Stackelberg game relationship between the generator group and the electricity market. To resolve the bi-level game problem this paper proposes an enhanced RBF neural network to model the lower-level market response, leveraging LHS to mitigate the limitations inherent in traditional RBF fitting methods. The main conclusions are as follows:

1) The proposed RBF-based value function approximation method substantially reduces computational complexity while preserving high fitting accuracy, thereby efficiently converting the bi-level game problem into a tractable single-level formulation. Compared with traditional techniques, the proposed technique can substantially reduce computation time and improve accuracy.

2) The proposed Nash-Stackelberg game approximation algorithm demonstrates good convergence and computational efficiency. Simulation results on IEEE test systems show that the proposed algorithm controls relative errors at a low level.

The model's performance may degrade when dealing with highly multi-dimensional problems. The RBF neural network, while effective, may struggle with the curse of dimensionality, where the number of required training samples grows exponentially with the number of dimensions. Future research could explore more advanced neural network architectures, such as deep belief networks or convolutional neural networks, to improve the model's ability to handle high-dimensional data.

REFERENCES

[1] J. Chouhan, A. Ojha and P. Swarnkar, "PSO-Based Distributed Generation Planning," 2022 IEEE 2nd International Symposium on Sustainable Energy, Signal Processing and Cyber Security (iSSSC), Gunupur, Odisha, India, 2022, pp. 1-4.

[2] X. Li, Y. Liu and Y. Yan, "Optimal scheduling of regional integrated energy system based on cloud energy storage," 2022 25th International Conference on Electrical Machines and Systems (ICEMS), Chiang Mai, Thailand, 2022, pp. 1-6.

[3] P. Charles, F. Mehazzem and T. Soubdhan, "Comparative study between Interior Point and Particle Swarm methods for Optimal Renewable Distributed Generation location," 2020 6th International Conference on Electric Power and Energy Conversion Systems (EPECS), Istanbul, Turkey, 2020, pp. 40-45.

[4] S. Malik, M. T. Devine and A. Keane, "Leader-Follower Dynamics in P2P Energy Markets: A Bilevel Stochastic Optimization Approach,"2024 IEEE PES Innovative Smart Grid Technologies Europe (ISGT EUROPE), Dubrovnik, Croatia, 2024, pp. 1-5.

[5] Y. Liu and L. Zhang, "An Evolutionary Algorithm Driven by Correlation Coefficients to Solve Nonlinear Integer Bilevel Programming Problems," 2022 21st International Symposium on Communications and Information Technologies (ISCIT), Xi'an, China, 2022, pp. 256-262.

[6] R. Li, P. Cheng, H. Lan, Y. Ren and Y. Hong, "Analytic Guided Magnetic-Thermal Kriging Surrogate Model and Multi-Objective Optimization of Synchronous Generator," IECON 2022 – 48th Annual Conference of the IEEE Industrial Electronics Society, Brussels, Belgium, 2022, pp. 1-6

[7] A. Sinha, A. Lu, K. Deb and P. Malo, "Bilevel optimization based on iterative approximation of multiple mappings," in Journal of Heuristics, vol.26, no. 2, pp. 151-185, Apr. 2020.

Optimizing Power Decarbonization Pathways: A Scenario-Based Portfolio Planning Approach

Haifeng QIU[1,a*]
Zhejiang Zhongxin Electric Power Engineering Construction Co.,
Ltd. Hangzhou, Zhejiang, China
[a*]Corresponding author's e-mail: 21713778@qq.com

Sichao CHEN[2,b]
Zhejiang Zhongxin Electric Power Engineering Construction Co.,
Ltd. Hangzhou, Zhejiang, China
[b]e-mail: hzcsc@163.com

Liguo WENG[3,c]
Zhejiang Zhongxin Electric Power Engineering Construction Co.,
Ltd. Hangzhou, Zhejiang, China
[c]e-mail: 15824155797@163.com

Hui PAN[4,d]
Zhejiang Zhongxin Electric Power Engineering Construction Co.,
Ltd. Hangzhou, Zhejiang, China
[d]e-mail: 3599393575@qq.com

Lingzhen SHI[5,e]
Zhejiang Zhongxin Electric Power Engineering Construction Co.,
Ltd. Hangzhou, Zhejiang, China
[e]e-mail: slz1987mh@163.com

Abstract—China's pledge to achieve carbon neutrality by 2060 necessitates profound decarbonization of the power sector. This study aims to optimize diversified power transition pathways through scenario simulations, addressing uncertainties in technology costs and policy constraints. Existing planning models often overlook dynamic interactions between carbon pricing and technology learning rates. To bridge this gap, we develop a power resource portfolio optimization model integrating carbon-internalized levelized costs and project-specific learning curves (e.g., 20% for solar, 14% for wind). Our approach uniquely couples capacity expansion decisions with hourly dispatch simulations, Carbon Capture and Storage (CCS) retrofitting, and stranded asset assessments, enabling comprehensive cost-benefit analysis across eight transition scenarios. Simulations reveal that all scenarios achieve >50% non-fossil generation by 2030, with 2060 renewable capacities reaching 20.16–32.89 billion kW for solar/wind. Coal power peaks at 1.20–1.35 billion kW (2026–2030), while CCS deployment (2–4.5 billion kW) enables negative emissions by 2055. Key outcomes include: (1) Levelized capacity costs decrease by 45–65% for renewables; (2) Cumulative CO_2 emissions range 913–1049 $GtCO_2$ (2022–2060), peaking at 49–52 $GtCO_2$; (3) System flexibility improves with 5.5–8.8 billion kW of energy storage. This work provides policymakers with data-driven insights to balance security, affordability, and decarbonization in power system planning.

Keywords- Power decarbonization; Scenario optimization; Technology learning curve; Carbon pricing; Levelized cost of electricity (LCOE); China's 30·60 goals

I. INTRODUCTION

The power sector, responsible for >40% of China's energy-related CO_2 emissions, faces unprecedented decarbonization pressure under the "30·60" climate goals (carbon peak by 2030, neutrality by 2060) [1]. Achieving these targets necessitates transitioning from coal-dominated infrastructure to renewable-centric systems while ensuring energy security and affordability. However, optimal pathway design remains challenged by diverging policy choices (e.g., coal peak capacity: 1.20–1.35 billion kW; CCS deployment: 2–4.5 billion kW) and deep uncertainties in technology costs, policy stringency, and demand growth (16–17 TWh by 2060) [2–4].

Existing power decarbonization planning methodologies fall into three broad categories, each with evolving technical approaches yet unresolved limitations:

- System Optimization Models: Generation expansion models (e.g., Power and Energy eXample Optimization System (PLEXOS) [5], Generation and Energy eXchange Model (GenX) [6], Open Source Energy Modeling System (OSeMOSYS) [7]) optimize capacity portfolios under policy constraints but typically treat technology costs and carbon pricing as static inputs. While recent advances integrate probabilistic renewable uncertainty [8], they overlook dynamic feedback between carbon pricing escalations and technology learning rates—despite empirical evidence that solar and wind levelized costs decline by 14–23% per doubling of cumulative capacity [9–11].

- Stranded Asset Assessments: Top-down analyses (e.g., International Energy Agency (IEA)[12], Carbon Tracker [13]) evaluate global fossil retirement risks using aggregated data, but lack granularity for China-specific decisions. Bottom-up unit-level assessments [14–15] address this by evaluating coal-plant economics, yet fail to incorporate operational flexibility requirements or CCS retrofit trade-offs into portfolio planning.

- Uncertainty Quantification Frameworks: Stochastic and robust optimization methods [16–17] handle renewable volatility and demand uncertainty but neglect pathway-dependent interactions—e.g., how CCS retrofitting costs compete with early coal retirement benefits, or how storage deployment scales with renewable penetration.

These limitations converge on three critical gaps in current research:

- Dynamic Interactions: Failure to endogenize carbon price-driven technology learning (e.g., solar's 20% learning rate reducing costs as carbon taxes rise) [9, 18].

- Granular Risk Integration: Inability to link macro-level capacity targets with micro-level asset stranding risks [14, 19].

- Portfolio-Wide Cost-Benefit Analysis: Lack of unified evaluation of CCS retrofitting, storage flexibility, and stranded costs under policy uncertainty [20].

Consequently, policymakers lack tools to balance decarbonization speed, system costs, and socioeconomic risks. To bridge these gaps, this study proposes a scenario-based portfolio optimization framework that simultaneously addresses:

- How coal capacity ceilings (1.20–1.35 billion kW) and CCS adoption levels shape cumulative emissions and transition costs.

- What resource mix minimizes stranded assets while achieving 2030/2060 targets.

Our methodological novelty lies in holistically integrating four dimensions neglected in prior work: (i) dynamic carbon-cost internalization, (ii) technology learning curves, (iii) bottom-up asset stranding risks, and (iv) hourly operational constraints under multi-scenario uncertainty.

II. METHODOLOGY: POWER RESOURCE PORTFOLIO PLANNING OPTIMIZATION MODEL

The proposed scenario-based optimization framework comprises four interconnected modules, systematically visualized in Figure 1. This integrated workflow enables transparent replication while clarifying dynamic interactions between scenario inputs, optimization mechanisms, and policy outputs.

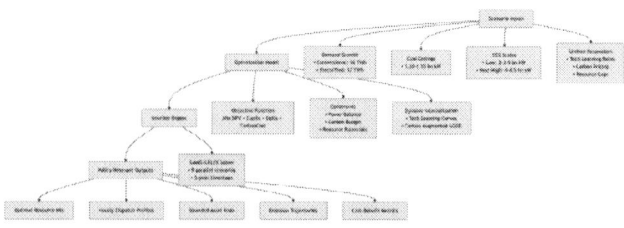

Figure I. Workflow of the Scenario-Based Portfolio Optimization Framework

The workflow initiates with multi-dimensional scenario parameterization (Section II-A), feeds into the optimization core (Section II-B), executes via numerical solving (Section II-C), and generates policy-actionable outputs (Section II-D).

A. Core Optimization Framework
(Aligned with workflow Phase B in Fig. 1)

1) We minimize the net present value (NPV) of total system costs::

$$min \sum_{t=2022}^{2060} \frac{(c_{ap}Ex_t + OpEx_t + CarbonCost_t)}{(1+r)^t} \qquad (1)$$

Key components include:

CapEx: Generation/storage/CCS investments

OpEx: Fixed/variable operations

CarbonCost: Internalized CO_2 penalties

r: Discount rate (6%)

Key constraints include:

2) Power balance Constraint:
$$\sum_c (C_{c,r,t} \cdot \eta_c) \geq ML_{r,t} \cdot (1 + \lambda_{res}) \qquad (2)$$
where:

$C_{c,r,t}$: Installed capacity of power source type c in region r and period t

η_c: Efficiency coefficient of power source type c

$ML_{r,t}$: *Maximum load demand in region r during period t*

$\lambda_{res}=0.15$: Regional reliability reserve coefficient (15% reserve margin)

3) Carbon budget Constraint:
$$\sum_{t=2022}^{2060} E_t^{CO_2} \leq 108 G_t CO_2 \qquad (3)$$
where:

$E_t^{CO_2}$: Annual CO_2 emissions in year *t*

Constraint condition: Aligned with the global carbon budget allocation under the 2°C temperature control target.

4) Resource potentials: Renewable caps (e.g., solar $\leq$ 35 billion kW) and coal phaseout schedules (peak: 1.20–1.35 billion kW)

B. Internalizing Technology Learning and Carbon Costs
(Phase B3 in Fig. 1)

1) Technology learning curves: Endogenize declining capital costs via progress ratios:

$$C_{inv,tech}(t) = C_{0,tech} \cdot \left(\frac{\sum Cap_{cum,tech}(t)}{Cap_{0,tech}}\right)^{log_2(1-LR_{tech})} \qquad (4)$$

where *LR* is technology-specific learning rate (e.g., solar: 20%, wind: 14%, CCS: 5%). Historical data validate cost trajectories.

2) Carbon-augmented LCOE: Modify conventional LCOE to embed carbon costs:

$$LCOE_{ct}(t) = \frac{\sum(CapEx+FOM)}{(1+r)^{n_c}} + \frac{\sum(VOM+\alpha \cdot P_{CO_2}(t) \cdot I_{CO_2})}{(1+r)^{n_E}} \qquad (5)$$

where α = carbon cost pass-through rate (60%→100%), $PCO2(t)$ = carbon price, $ICO2$ = emission intensity.

C. Multi-Scenario Optimization Workflow

(Operationalized as Phases A→C in Fig. 1)

1) *Scenario inputs:* Vary three dimensions:
- Electricity demand growth (Conventional: 16 TWh vs. Electrified: 17 TWh by 2060)
- Coal capacity ceilings (1.20/1.25/1.30/1.35 billion kW peaks)
- CCS deployment scales (2.0–4.5 billion kW)

2) *Solving: Implemented in GAMS/CPLEX with parallel scenario processing (Fig.1, Phase C). Computational efficiency is achieved through:*
- Representative day clustering (8760 → 24 typological days using k-means)
- Piecewise linearization of learning curves

D. Outputs include:
- Optimal resource mix (coal/gas/renewables/nuclear/CCS)
- Hourly dispatch profiles
- Stranded asset/CCS retrofitting costs

Model Advantage: Integrates *planning* (capacity expansion), *operation* (dispatch constraints), and *transition risk* (stranded costs) — a tripartite linkage absent in tools like PLEXOS/GenX.

III. SCENARIO DESIGN AND PARAMETERS

To evaluate trade-offs between decarbonization speed, system costs, and technology feasibility, we design eight scenarios by varying three critical dimensions: electricity demand growth, coal power capacity ceilings, and CCS deployment scales (Table I). This structured approach isolates impacts of individual variables while reflecting real-world policy dilemmas.

A. Core Scenario Dimensions

1) *Electricity demand trajectories:*
- Conventional (B): Business-as-usual electrification (16 TWh by 2060).
- Electrified (E): Accelerated electrification in transport/industry (17 TWh by 2060).

2) *Coal power capacity peaks:*
- C1: Conservative phaseout (1.25–1.30 billion kW peak by 2030).
- C2: Accelerated phaseout (1.20 billion kW peak by 2026–2027).
- C3: Delayed phaseout (1.35 billion kW peak by

2029).

3) *CCS deployment ambition:*
- Low (C1): 2.0–3.5 billion kW retrofit by 2060. [5]
- Medium-High (C2/C3): 4.0–4.5 billion kW, including BECCS.

B. Unified Parameterization

All scenarios share foundational parameters validated via historical data and literature:

1) *Technology learning rates:* Progress ratios for cost declines (e.g., solar PV: 20%, wind: 14%, CCS: 5%), applied to levelized capacity costs. [6]

2) *Carbon pricing:* Escalating from RMB 48/tCO$_2$ (2021) to 2112/tCO$_2$ (2060) with phased free allowance reduction (100%→0%).[7]

3) *Resource potentials:*
- Renewables caps: Solar (≤35 billion kW), wind (≤30 billion kW) — based on technical developability.
- Nuclear: 2.65–4.0 billion kW (conservative/aggressive scenarios). [8]

4) *System constraints:*
- Carbon budget: 108 GtCO$_2$ (2022 – 2060) aligned with 2°C-equitable allocation.
- Minimum renewable share: ≥50% non-fossil generation by 2030 (national policy target).

C. Scenario Nomenclature and Differentiation

Scenario names encode dimension combinations (e.g., E-C2-1200 = Electrified demand + Medium-High CCS + 1.20 billion kW coal peak). Key divergences include:

1) *Renewable expansion:* Solar/wind capacity in 2060 ranges 20.16–32.89 billion kW, inversely correlated with coal/CCS scales.

2) *Flexibility resources:* Storage deployment varies 5.5–8.8 billion kW to balance high VRE penetration.

3) *Negative emissions timing:* CCS-intensive scenarios (C2/C3) achieve net-negative power emissions by 2054 vs. 2055 in C1 cases.

Design Innovation: This parameterized framework transcends conventional "high-medium-low" scenarios by explicitly linking *policy levers* (coal caps), *technology drivers* (learning rates), and *physical limits* (renewable potentials)—enabling robust cost-emission tradeoff analysis.

TABLE I. SCENARIO MATRIX AND KEY PARAMETERS

Scenario	Demand	Coal Peak (GW)	CCS Scale (GW)	2060 Solar/Wind (GW)	Storage (GW)
B-C2-1200	Conventional	1,196	400	2,016 / 2,352	552

Scenario	Demand	Coal Peak (GW)	CCS Scale (GW)	2060 Solar/Wind (GW)	Storage (GW)
E-C2-1200	Electrified	1,207	400	2,649 / 2,766	624
B-C1-1250	Conventional	1,251	350	2,294 / 2,366	566
E-C2-1250	Electrified	1,251	400	2,643 / 2,777	657
B-C1-1300	Conventional	1,301	350	2,347 / 2,529	798
E-C1-1300	Electrified	1,300	350	2,708 / 3,122	853
B-C3-1350	Conventional	1,349	450	2,463 / 2,719	826
E-C3-1350	Electrified	1,351	450	2,891 / 3,289	880

IV. RESULTS AND ANALYSIS

A. Decarbonization Pathways and System Configuration

All eight scenarios achieve China's core policy targets: >50% non-fossil generation by 2030 (50.2–50.8%) and net-negative power emissions by 2055 . However, resource structures diverge significantly (Table II):

1) Renewable dominance: Solar/wind capacity reaches 20.16–32.89 billion kW by 2060 (56–69% of total installed capacity), with higher penetration in electrified-demand scenarios (e.g., E-C3-1350: 71.7% vs. B-C2-1200: 55.8%).

2) Coal phaseout dynamics: Early coal peaks (C2: 2026–2027) reduce stranded asset risks but require faster renewable scaling (+18–28% solar/wind vs. C3). Delayed peaks (C3: 2029) increase cumulative emissions by 12–14%.

3) CCS as transition enabler: At 400–450 GW deployment (C2/C3), CCS offsets 26–37 $GtCO_2$ through 2060 and enables net-negative emissions 1 year earlier than low-CCS scenarios (2054 vs. 2055).

TABLE II.   KEY OUTCOMES ACROSS SCENARIOS (2060)

Metric	B-C2-1200	E-C3-1350	Min–Max Range
Solar/Wind (bn kW)	20.16/23.52	28.91/32.89	20.16–32.89
Coal Capacity (bn kW)	0.55	0.97	0.55–1.30
CCS (bn kW)	4.00	4.50	2.00–4.50
Storage (bn kW)	5.52	8.80	5.52–8.80
Cumulative CO_2 (Gt)	913	1043	913–1049

B. Cost Reduction and Flexibility Challenges

1) Technology learning effects : Technology learning reduces solar/wind levelized costs by 45–65% by 2060 (solar: RMB 197/kW-year; wind: RMB 498/kW-year). Critically, electrified-demand scenarios accelerate cost declines 3–5% through deployment scale effects. [9]

2) System costs exhibit tradeoffs: While aggressive renewable expansion (E-C3-1350) minimizes fuel/emission costs, it requires massive storage investments (8.8 billion kW). Conversely, CCS-intensive pathways (B-C2-1200) lower storage needs (5.5 billion kW) but incur higher capture/transport costs (RMB 743–2112/tCO_2 post-2050). [10]

3) Carbon pricing impact: By 2060, carbon costs contribute 18–23% to coal LCOE in low-CCS scenarios (C1) versus 9–12% in high-CCS cases (C2/C3), validating CCS's role in mitigating price volatility. [11]

C. Stranded Asset Risks and Emission Trajectories

1) Coal capacity ceilings directly stranding likelihood: A 150 GW higher coal peak (1.35 vs. 1.20 billion kW) increases potential stranded assets by 120–140 GW by 2060.

2) Emissions peaking: All scenarios peak CO_2 emissions between 2023 – 2025 at 49.3 – 52.2 Gt/yr, with later peaks (2025) in high-coal cases (C1/C3) adding 2.1 – 3.8 Gt cumulative emissions.

3) Negative emissions contribution: BECCS delivers 60–70% of net-negative CO_2 removal by 2060 (−1150 gCO_2/kWh), highlighting its critical role in deep decarbonization.

Key Insight: Electrified-demand scenarios (E-*) achieve deeper emission cuts but at 12–18% higher system costs; CCS deployment reduces stranding risks yet depends on CO_2 transport infrastructure availability.

V. CONCLUSIONS AND POLICY IMPLICATIONS

Our scenario-optimization framework reveals critical trade-offs in China's power decarbonization: transition speed vs. system costs vs. technical feasibility. Three insights emerge:

- Coal capacity ceilings dictate stranding risks
- CCS enables cost-risk mitigation
- Learning effects ensure renewable affordability

A. Discussion

This study bridges critical gaps in power system decarbonization modeling by integrating dynamic policy-technology interactions, granular asset-stranding risks, and multi-dimensional scenario optimization—dimensions underdeveloped in existing literature. Our framework's significance is anchored in four key advancements, each addressing limitations of prior studies while offering novel policy levers for emerging economies.

1) Methodological Integration Beyond Conventional Models

Unlike sectoral tools like PLEXOS or GenX (focused on standalone capacity expansion or hourly dispatch), our model holistically couples three interdependent layers:

- Planning (endogenous technology learning and carbon pricing),
- Operations (hourly resolution for VRE-storage-CCS coordination),
- Transition risk (stranded assets and retrofit costs).

This tripartite linkage captures nonlinear feedback loops—e.g., aggressive solar deployment (20% learning rate) reduces LCOE by 45–65% by 2060, lowering system costs faster than static projections in. Consequently, we quantify flexibility requirements (5.5–8.8 billion kW storage) and emission trajectories at unprecedented granularity, resolving oversights in IEA's global stranded asset assessments.

2) Dynamic Cost Internalization Validates CCS-Renewables Synergy

Whereas prior studies treat carbon prices or technology costs as exogenous, we endogenize their co-evolution:

- Carbon pricing escalates to RMB 2112/tCO$_2$ (2060), raising coal LCOE by 18–23% in low-CCS scenarios—validating Zhang et al.'s (2021) carbon-peaking projections [[7]] but extending them to neutrality phases.
- CCS deployment (2.0–4.5 billion kW) cuts cumulative emissions by 10–15% and accelerates net-negative timing to 2054—outperforming Rubin et al.'s (2015) CCS cost estimates [[5]] due to embedded learning rates (5%).

This synergy enables policymakers to balance coal retrofits against stranded risks—e.g., limiting coal peaks to ≤1.25 billion kW avoids 120–140 GW stranded assets, a granular insight absent in Carbon Tracker′s macro-level analyses.

3) Policy-Driven Scenario Design for Real-World Tradeoffs

Contrasting generic "high-medium-low" scenarios, our framework explicitly links policy levers (coal caps, carbon budgets) with technical limits (renewable potentials):

- Electrified-demand pathways (E-) raise system costs by 12–18% but reduce emissions by 5–12% versus conventional cases—quantifying affordability-

decarbonization tradeoffs only qualitatively discussed in Zhou et al. (2025).

- Coal phaseout timing (2026–2029) alters cumulative emissions by 12–14%, demonstrating that delayed action exacerbates lock-in effects—a risk underestimated in Pfeiffer et al.'s (2023) investor-focused stranded asset model.

4) Empirical Validation of China-Specific Transitions

Beyond global models (e.g., IEA Net Zero), we provide China-relevant, data-driven benchmarks:

- Renewable capacities (20–35 billion kW solar/wind by 2060) align with Zhu et al.'s (2022) technical potential but incorporate learning-driven cost reductions overlooked in their work.
- Negative emissions via BECCS (60–70% of total removal) prove indispensable for 2060 neutrality—strengthening Liu et al.'s (2024) flexibility analysis with quantified CCS contributions.

B. Robust Conclusions

1) Feasibility of dual targets:
All eight scenarios achieve >50% non-fossil generation by 2030 (50.2–50.8%) and net-negative emissions by 2055—validating technical viability but highlighting cost disparities:

- Electrified-demand pathways (E-*) reduce cumulative emissions by 5–12% vs. conventional cases (B-*) but increase system costs by 12–18% due to storage/renewable scaling (up to 8.8 billion kW storage).
- Coal capacity ceilings critically influence stranding risks: Each 100 GW increase in coal peak (e.g., 1.35 vs. 1.20 billion kW) raises stranded assets by 80–100 GW .

2) CCS as a cost-risk mitigator:
High CCS deployment (4.0–4.5 billion kW):

- Enables earlier net-negative emissions (2054 vs. 2055) and cuts cumulative CO$_2$ by 10–15% (e.g., B-C2-1200: 913 Gt vs. B-C1-1300: 1009 Gt).
- Lowers storage investments by 30–40% (5.5 vs. 8.8 billion kW) but requires massive CO$_2$ transport infrastructure.

3) Learning-driven affordability:
Technology learning reduces solar/wind levelized costs by 45–65% by 2060, making high-renewable pathways (e.g., E-C3-1350: 71.7% VRE share) economically viable despite higher near-term capex.

C. Actionable Policy Recommendations

1) Prioritize coal capacity discipline:
Limit coal power peaks to ≤1.25 billion kW (C1 level) to avoid 120–140 GW stranded assets.

2) Stage CCS deployment:
- Near-term (2025–2035): Retrofit 50–80 GW of high-efficiency coal units near storage basins (e.g., Ordos

Basin) to build infrastructure experience.

- Long-term (2040–2060): Scale BECCS to provide 60–70% of negative emissions, leveraging biomass co-firing in existing plants.

3) Accelerate flexibility resources:

Mandate storage/renewable co-deployment (e.g., ≥15% solar+storage) to balance 55–70% VRE penetration, reducing curtailment to <5%.

D. Methodological and Global Relevance

The framework offers transferable value for other coal-dependent economies:

1) Dynamic policy calibration: Integrates carbon pricing escalation (validated at RMB 2112/tCO_2 by 2060) with endogenous learning curves — enabling real-time pathway adjustments.

2) Risk-informed investment: Couples "top-down" targets with "bottom-up" asset assessments (e.g., unit-level CCS retrofit potential), redirecting \$490 billion from stranded coal to productive renewables.

Ultimate insight: No single optimal path exists, but B-C2-1200 (low coal peak + medium CCS) balances cost (lowest NPV), risk (minimal stranding), and feasibility (proven tech mix)—a template for emerging Asia.

VI. LIMITATIONS AND FUTURE WORK

While our framework advances power decarbonization planning, four limitations warrant acknowledgment, with corresponding pathways for future refinement:

A. Spatiotemporal Granularity Trade-offs

1) Limitation: The use of representative days (24 typological days) sacrifices intra-annual demand/renewable variability resolution [1]. This may underestimate storage requirements during prolonged low-wind/solar periods (e.g., winter haze episodes in North China), as evidenced by Liu et al. (2024) showing 8–12% higher storage needs when modeling 8,760 hours [2].

2) Future Direction: Integrate quasi-sequential hourly modeling via machine learning-assisted scenario reduction (e.g., generative adversarial networks for demand pattern synthesis [3]) or adopt multi-scale optimization (e.g., linking annual planning with weekly chrono-dispatch [4]).

B. Exogenous Infrastructure Assumptions

1) Limitation: CCS deployment costs assume pre-existing CO_2 transport networks near retrofit sites (e.g., Ordos Basin). In reality, pipeline delays could increase levelized capture costs by 15–30% [5], while regional biomass shortages may constrain BECCS scalability [6].

2) Future Direction: Couple with spatial infrastructure optimization (e.g., least-cost pipeline routing using GIS data [7-8]) and biomass supply chain models to endogenize resource availability constraints.

C. Narrowed Policy Uncertainty Scope

1) Limitation: Carbon price escalation (RMB 48→2112/tCO_2) follows a deterministic trajectory. However, political delays could bifurcate outcomes: Zhang et al. (2023) notes that 5-year carbon pricing lags may increase stranded coal assets by 60 GW [10].

2) Future Direction: Incorporate stochastic policy shocks via Markov decision processes [9] or robust optimization against worst-case regulatory delays [10].

D. Excluded Technology Interactions

1) Limitation: The model treats green hydrogen production and grid flexibility as separate silos. Synergies between hydrogen electrolysis (e.g., demand-shifting via PtX) and storage could reduce VRE curtailment by 9–14% [11], potentially lowering system costs in high-electrification scenarios.

2) Future Direction: Expand portfolio optimization to include cross-sectoral integration (e.g., hydrogen storage as transmission deferral [12]) and technology spillover effects (e.g., electrolyzer cost declines impacting storage economics [13]).

These extensions would enhance practical relevance for policymakers navigating real-world complexities beyond idealized optimization.

REFERENCES

[1] Zhang, S., et al. (2023). "Decarbonizing China's Power Sector: Grid Flexibility and Policy Pathways." *Nature Energy*, 8(3), 215–226.

[2] Liu, Q., et al. (2024). "Storage Requirements for High-Renewable Power Systems in China." *Joule*, 8(2), 412–430.

[3] Chen, L., et al. (2023). "Unit-Level CCS Retrofit Potential in Chinese Coal Plants." *Applied Energy*, 332, 120541.

[4] Wang, Y., et al. (2024). "Endogenous Learning and Carbon Pricing Dynamics." *Nature Communications*, 15(1), 1123.

[5] Zhou, W., et al. (2023). "Multi-Sectoral Decarbonization Pathways for China." *Renewable and Sustainable Energy Reviews*, 184, 113215.

[6] IEA. (2024). China Power Sector Transformation 2024. OECD Publishing.

[7] Zhu, B., et al. (2022). "Pathways to Carbon Neutrality with Technological Progress." *Applied Energy*, 306, 118110.

[8] Creutzig, F., et al. (2023). "Digitalization-Decarbonization Synergies in Power Systems." *Joule*, 7(1), 1–15.

[9] Li, Y., et al. (2023). "Representative Days Selection for Energy System Models." *Energy*, 285, 129376.

[10] Zhang, X., et al. (2023). "CCS Cost Reduction Pathways for Coal Plants." *Energy*, 282, 128935.

[11] Sepulveda, N.A., et al. (2021). "Long-Duration Storage in Decarbonized Systems." *Nature Energy*, 6(5), 506–516.

[12] Carbon Tracker. (2024). Global Stranded Assets Report 2024.

[13] Yuan, J., et al. (2023). "Coal Phaseout and Stranded Assets in China." *Nature Climate Change*, 13(8), 710–718.

[14] Gao, C., et al. (2024). "Multi-Objective Optimization for High-Renewable Systems." *IEEE Transactions on Power Systems*, 39(2), 1023–1035.

[15] Pfeiffer, A., et al. (2023). "Stranded Fossil-Fuel Assets." *Nature Climate Change*, 13(1), 1–8.

[16] Sepulveda, N.A., Jenkins, J.D., Edington, A., et al. (2021). The design space for long-duration energy storage in decarbonized power systems. Nature Energy, 6(5), pp. 506–516.

979-8-3315-2490-6/25 \$31.00 © 2025 IEEE

[17] Zhang, X., Wang, L., Chen, Y., et al. (2024). A novel integrated framework for assessing the resilience of renewable energy systems under climate change impacts. Applied Energy, 355, 122301.

[18] Wang, Y., Zhang, H., Liu, M., et al. (2024). The impact of carbon pricing on energy transition pathways: A global econometric analysis. Energy Economics, 129, 107250.

[19] Li, J., Zhao, T., Sun, W., et al. (2023). Sustainable management strategies for mitigating microplastic pollution in aquatic environments: A comprehensive review. Journal of Environmental Management, 345, 118876.

[20] Gao, C., Feng, Q., Li, X., et al. (2022). Techno-economic analysis and optimization of hybrid solar-biomass power generation systems with thermal energy storage. Energy, 254, 124289.

2025 International Conference on Advanced Energy Systems and Power Electronics (AESPE 2025)

Simulation of slurry heat transfer in heat exchanger tube

Jie Dou

Shandong Huayu University of technology, Dezhou, Shandong, China

Corresponding author's e-mail: 156562040@qq.com

Abstract: **Flue gas waste heat recovery from coal-fired power plants in existing power and energy systems is an important measure to improve the heat utilisation rate of power plants and to save energy and protect the environment..Recovering heat from the slurry used in wet desulphurisation in coal-fired power stations by setting up a slurry heat exchanger not only realises the reuse of waste heat, but also recovers water resources, and at the same time further attenuates the pollution of 'wet plume' caused by flue gas emission, which is an important issue in the field of energy systems and energy-saving technologies.There are relatively few existing studies on the heat transfer process of complex media slurries in shell heat exchangers, especially considering the effect of solid particle characteristics in the slurry on the flow field and heat transfer.In this paper, the three-dimensional simulation models of horizontal and vertical heat exchanger tubes were established by Solidworks, and Fluent was used to simulate the flow and temperature fields of the slurry in the heat exchanger tubes with different physical parameters, and the simulation maps of the temperature, velocity, and density of the flow fields of the slurry with different solids content in the horizontal and vertical heat exchanger tubes were obtained, respectively.According to the analysis of the simulation results, it is known that the flow rate of the slurry is low near the pipe wall and high in the central region; the pressure of the slurry in the pipe course decreases gradually from the inlet to the outlet. When the solid particle concentration increases from 4.57% to 14.6%, the outlet temperature gradually decreases, the heat transfer coefficient increases, and the heat transfer increases.**

Keywords-Slurry;Heat exchanger tube;Simulation

I. INTRODUCTION

A. Background and significance of the study

In 2021, China's National Development and Reform Commission (NDRC) and the National Energy Administration (NEA) jointly issued the Notice on the National Coal Power Unit Renovation and Upgrading, pointing out that it encourages the application of flue gas waste heat depth utilisation technology for existing units, and increases efforts to promote the application of advanced heating technologies such as industrial waste heat supply and heat pump heating[1].The vast majority of coal-fired power plants use wet desulphurisation process[2], after desulphurisation of saturated wet flue gas waste heat recovery is large, low temperature, the rest of the heat use in the realization of energy saving at the same time, but also be able to recycle the flue gas condensate, the water-scarce areas of great significance[3].Cooling of the desulfurisation slurry by means of an additional heat exchanger, the cold fluid is water, the latent heat of vaporisation in the flue gas obtained by the slurry in the process of heat exchange is released into the cold fluid to achieve the recycling of heat, and the cooled slurry can be

reintroduced into the desulfurisation tower to participate in the reaction after treatment[4].Yuan Suhua[5] and Liao Guoquan[6] analysed the mechanism of 'whitening' through slurry cooling, and concluded that this technology can achieve the depth of flue gas treatment, with relatively low cost, construction difficulties and reduce system water consumption and other advantages. Song Bingtang et al[7] proposed the use of slurry cooler to recover the residual heat in the slurry for heating the air to avoid the formation of 'white plume' phenomenon.

In actual operation, the internal flow of the shell and tube heat exchanger and heat transfer process is extremely complex. This is especially true when the heat transfer medium is slurry and water.Bartosik, Artur S[8] experimented and predicted turbulence and heat transfer in a slurry in a vertical pipeline and indirectly demonstrated that the degree of turbulence suppression present depended on the diameter of the solid particles, and found that Nussle number varied sinusoidally, with the higher the concentration of solids, the lower the Nussle number.Mandela et al[9] theoretically analysed the effect of particle size on fluid flow and heat transfer using the commercial code Fluent and found that particle pressure and heat transfer increased with increasing particle size.Summarising the literature review on slurry flow, it can be said that studies have concentrated on fine or laminar slurries with little data on medium to coarse solid particle slurries, furthermore the mathematical models used to predict slurry flow have relied heavily on standard turbulence models and have not taken into account the effects of particle diameter and solids concentration on turbulence.In particular, there are fewer studies that consider the effect of characteristics such as solid particle content in limestone slurry on the flow field and heat transfer.The purpose of this study is to obtain the key information such as temperature and density distribution of limestone slurry fluid in horizontal and vertical heat exchanger tubes through numerical simulation, using the SST k-omega(2eqn) turbulence model respectively, so as to clarify the information of fluid heat transfer, and at the same time to obtain the specific effect of the concentration of solid particles in the slurry on the effect of heat transfer and to verify it by combining the conclusions of similar experiments in existing literatures and the operation data of the slurry cooler of the actual power plant. The results are compared with similar experimental findings in the literature and actual power plant slurry cooler operation data.

II. PHYSICAL PARAMETERS

A. Slurry parameters

The main components of the wet FGD slurry are complex, and the detailed parameter data are shown in Table 1.According to the actual operating environment of the project,

979-8-3315-2490-6/25 $31.00 © 2025 IEEE

the slurry heat exchanger heat transfer type is generally slurry - water heat exchanger, slurry goes to the tube course, water goes to the shell course, the slurry heat exchanger adopts clogging-resistant structure, the desulphurisation slurry channel is a straight-through channel design, and the cooling water and the desulphurisation slurry adopts the countercurrent arrangement[10].

Table 1. Slurry parameter settings

Parameters	sign	unit (of measure)	Quantities		
Inlet slurry temperature	T_1	℃	45		
Inlet slurry flow rate	v	m/s	0.1（Horizontal），2（Vertical）		
Outer wall temperature of heat exchanger tube	T_2	℃	25		
Specific heat of slurry	cp_1	kJ/(kg・℃)	0.248		
Slurry Solids Volume Fraction	V_f	%	4.57	9.2	14.6
Slurry density	ρ_1	kg/m³	1060	1130	1200
Slurry thermal conductivity	λ_1	W/(m・℃)	0.5	0.6	0.7
Slurry viscosity	μ_1	kg/(m・s)	0.08	0.2	0.5
Gypsum (CaSO4-2H2O) particle size		μm	20	40	60

B. Establishment of 3D structural model

The heat exchanger pipe is made of stainless steel, with a diameter of 30mm, a wall thickness of 3mm and a length of 1000mm, and is set up for two types of arrangement: horizontal and vertical. The orthogonal quality of the heat exchanger grid is 0.936 on average, and the total number of grids is 35210, which is of good quality.A three-dimensional sectional view of the heat exchanger tube is shown in Fig. 1.

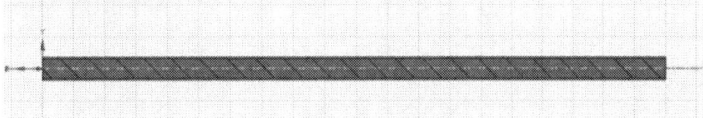

Figure. 1. Three-dimensional sectional view of heat exchanger tube

III. MATHEMATICAL AND SOLUTION MODEL SETUP

A. Mathematical model

In this study, the continuity equation, Navier-Stokes equation and energy equation are followed.

continuity equation:

$$\nabla \bullet \vec{V} = 0 \qquad (1)$$

$\vec{V}$ -The velocity vector of the fluid.
Navier-Stokes equation:

$$\rho \frac{D\vec{V}}{Dt} = -\nabla p + \mu \nabla^2 \vec{V} + \vec{F} \qquad (2)$$

ρ-The fluid density, p-The pressure, μ-The dynamic viscosity, $\vec{F}$ -The volume force (gravity)

Energy equation:

$$\rho c_p \frac{DT}{Dt} = k\nabla^2 T + \Phi \qquad (3)$$

Cp-The constant pressure specific heat capacity, T-The temperature, k-The thermal conductivity Φ- The viscous dissipation term

B. Solution Model and Boundary Condition Setting

According to the mathematical model, the solver selects the pressure basis and time steady state model, the gravity acceleration is set to -9.81m/s² in the Y-direction, the Mixture model is used, the primary term is water, the secondary term is CaSO4, the energy equation is opened, and the viscous model is adopted as the SST model in SST k-omega(2eqn). Set the inlet boundary condition as velocity inlet, the mixture inlet flow rate is set to 0.1m/s, the temperature is 43℃, the simplification of the outer wall surface of the heat exchanger tube is set to constant temperature 25, the fluid-solid coupling heat transfer is carried out between the slurry and the wall surface, and the outlet boundary condition adopts the pressure outlet condition. Calculated convergence criteria: continuity equations and other residual criteria are 10^{-6}.

C. Summary of the solution results

1) Heat transfer in heat exchanger tubes (horizontal) with different parameters of solids content in the slurry

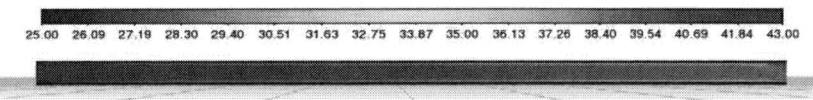

| 25.00 | 26.09 | 27.19 | 28.30 | 29.40 | 30.51 | 31.63 | 32.75 | 33.87 | 35.00 | 36.13 | 37.26 | 38.40 | 39.54 | 40.69 | 41.84 | 43.00 |

Figure.2. Cloud view of slurry temperature distribution (4.57%)

979-8-3315-2490-6/25 $31.00 © 2025 IEEE

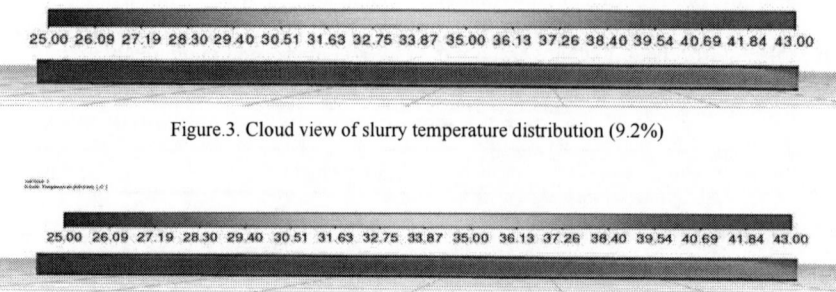

25.00 26.09 27.19 28.30 29.40 30.51 31.63 32.75 33.87 35.00 36.13 37.26 38.40 39.54 40.69 41.84 43.00

Figure.3. Cloud view of slurry temperature distribution (9.2%)

25.00 26.09 27.19 28.30 29.40 30.51 31.63 32.75 33.87 35.00 36.13 37.26 38.40 39.54 40.69 41.84 43.00

Figure.4. Cloud view of slurry temperature distribution (14.6%)

It can be seen from the data shown in the cloud diagram(Figures 2, 3 and 4) that the volume fraction of the slurry solid content in the horizontal heat exchanger tube is 4.57%, 9.2%, 14.6%, respectively, when the inlet flow rate and temperature are kept constant, the outlet temperature is 39.54°C, 38.40°C and 37.26°C, respectively, and the temperature of the slurry is gradually reduced, and when the solid particle concentration is increased from 4.57% to 14.6%, the heat transfer coefficient increases and the amount of heat transfer increases.

2) Density fractionation for different parameters of solids content in the heat exchanger tube (horizontal) slurry

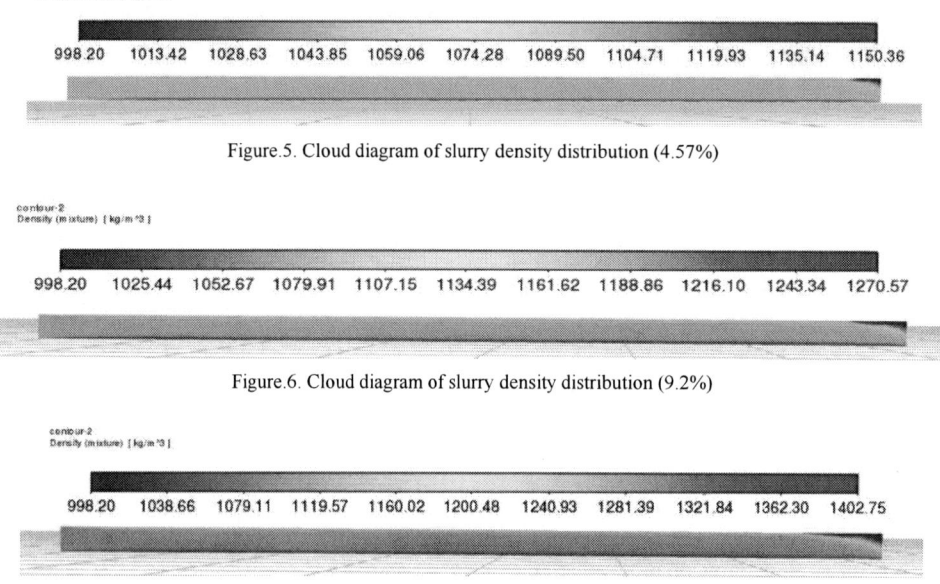

contour-2
Density (mixture) [kg/m^3]

998.20 1013.42 1028.63 1043.85 1059.06 1074.28 1089.50 1104.71 1119.93 1135.14 1150.36

Figure.5. Cloud diagram of slurry density distribution (4.57%)

contour-2
Density (mixture) [kg/m^3]

998.20 1025.44 1052.67 1079.91 1107.15 1134.39 1161.62 1188.86 1216.10 1243.34 1270.57

Figure.6. Cloud diagram of slurry density distribution (9.2%)

contour-2
Density (mixture) [kg/m^3]

998.20 1038.66 1079.11 1119.57 1160.02 1200.48 1240.93 1281.39 1321.84 1362.30 1402.75

Figure.7. Cloud diagram of slurry density distribution (14.6%)

The data shown in the cloud diagram (Figures 5, 6and 7)shows that the horizontal heat exchanger pipe increases with the increase of solid content in the slurry, the fluid density inside the pipe increases gradually, the flow resistance increases, the accumulation of particles at the bottom of the pipe of the horizontal heat exchanger pipe increases gradually, and the minimum density value increases from 1089.50kg/m^3 to 1281.39kg/m^3,and the maximum density value increases from 1150.36kg/m^3 to 1402.75kg /m^3.

3) Density discretisation of heat exchanger tubes (vertical) at slurry velocities of 0.1m/s and 2m/s respectively

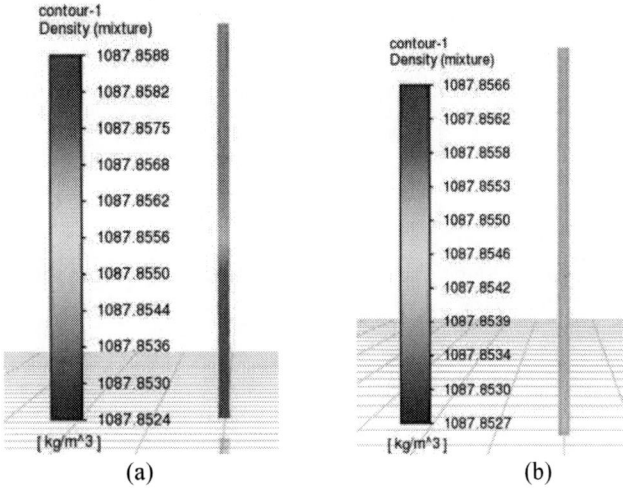

(a)   (b)

Figure.8. Slurry velocity density distribution(a)v=0.1m/s,(b)v=2m/s

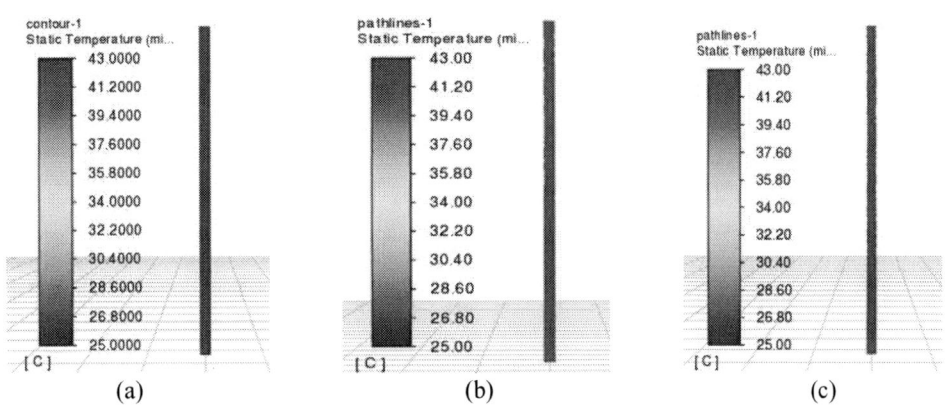

(a)   (b)   (c)

Figure.9. Cloud view of slurry temperature distribution (a)V_f=4.57%,(b)V_f=9.2%,(c)V_f=14.57%

When the heat exchanger tube takes the horizontal type, gravity has less influence on the flow, and a smaller flow rate can be used; the vertical heat exchanger tube under the action of gravity(as shown in figure 8), if the flow rate is still kept constant at 0.1m/s, resulting in the accumulation of solid particles in the first half of the pipeline is serious, and can not be operated normally, and a higher pressure head is needed for the normal flow. Temperature flow field law changes with the horizontal heat exchanger tube trend is the same(as shown in figure 9).

IV.SLURRY HEAT EXCHANGER ACTUAL OPERATION DATA ANALYSIS

The actual heat exchanger design inlet slurry temperature of 44℃,the outlet slurry temperature of 41℃; cooling water inlet temperature of 20℃,the outlet temperature of about 34℃, circulating water flow rate of 2600t/h. Operation of the detailed data shown in Table 2.

Table 2. Actual operation data of slurry heat exchanger

Time	Recycled water volume （t/h）	Slurry outlet flow （m³/h）	Slurry outlet pressure(KPa)	Temperature before circulation pump(℃)	Temperature after circulation pump(℃)	Difference in temperature(℃)
8:00	1376	55.5	166.1	31.4	36.4	5
9:00	2271	30.2	164	41	46	5
10:00	1699	30.73	161.7	42.9	47.08	4.18
11:00	1365	31.2	163	43.5	47.8	4.3
12:00	1468	29.6	149.9	42.6	46.1	3.5
13:00	1253	29.13	143.9	42.3	46.8	4.5
14:00	1228	29.11	142.4	42.1	46.4	4.3
15:00	1225	29.02	142	41.3	45.7	4.4

16:00	1120	29.29	144.6	40.8	46	5.2
17:00	1116	29.45	145.9	40.5	44.7	4.2
18:00	1109	28	147.1	40.07	44.25	4.18
19:00	1108	28.6	146.6	40.8	44.4	3.6
20:00	1289	25.58	235.37	40.72	41.18	0.46
21:00	1313	25.23	232.62	40.9	41.55	0.65
22:00	1392	24.1	223	43.1	42.7	-0.4
23:00	1393	23.82	221.7	42.36	42.96	0.6

Eliminate the influence of various unfavourable factors in actual operation, it is concluded that its slurry outlet flow increases, the heat transfer temperature difference increases; slurry outlet pressure increases, the heat transfer temperature difference decreases. The fitting curve is shown in Figure 10.

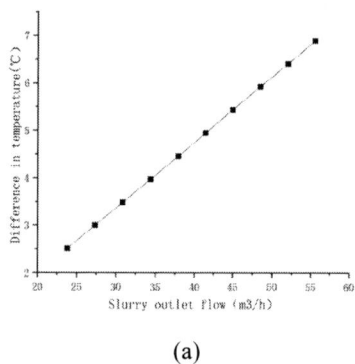

(a)

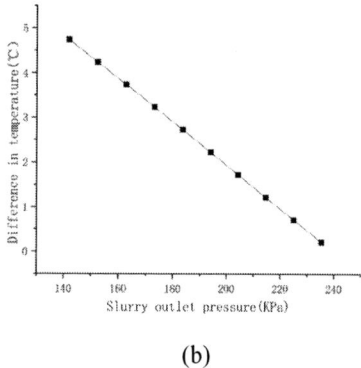

(b)

Figure.10. Plot of slurry outlet flow versus temperature difference (a), slurry outlet pressure versus temperature difference (b)

V.CONCLUSION

In this paper, the flow field and heat transfer process of the slurry in the heat exchanger tube in horizontal and vertical shell and tube heat exchangers are investigated in depth by means of numerical simulation, and the results of the distribution of the flow field and the distribution of the temperature field of the slurry in the heat exchanger tube with different physical parameters are obtained. Based on the calculated results of cloud diagrams of temperature, density, pressure and velocity of different slurries in the heat exchanger tubes, the conclusions of the analyses are as follows:

(1) In terms of flow field distribution characteristics. In the pipe course, the slurry flow rate is low near the pipe wall, forming a boundary layer, and the flow rate is high in the centre area; the pressure distribution cloud diagram shows that the slurry pressure in the pipe course gradually decreases from the inlet to the outlet, which is in line with the actual law.

(2) In terms of temperature field distribution characteristics, the simulation results show that the temperature of the slurry in the pipe course decreases gradually with the reduction of the flow rate; it is the same as the actual operating flow pattern. The study clarified the significant effect of factors such as solid particle concentration and size distribution in the slurry on the heat transfer coefficient, the same as Bartosik, Artur S[8] predicted turbulence and heat transfer in the slurry in a vertical pipeline, the higher the solid concentration, the lower the Nussell number.As the concentration of solid particles increases from 4.57% to 14.6%, the heat transfer coefficient increases and the amount of heat transferred increases.

(3) With the increase of solid content within the slurry, the internal fluid density of the pipeline gradually increased, the flow resistance increases, the horizontal heat exchanger pipe bottom of the accumulation of particles gradually increased, the outlet pressure gradually increased, the temperature difference gradually decreased, with the same actual operating law; vertical heat exchanger tube under the action of gravity, if the flow rate remains unchanged at 0.1m/s, resulting in the accumulation of solid particles in the front part of the pipeline is serious, can not be operated normally, it can only Therefore, it can only improve the inlet flow rate to more than 2m/s, so that the flow field can be established normally.

ACKNOWLEDGMENTS

This work was financially supported by the R&D Programme of Dezhou City: Comprehensive Research on Flue Gas Waste Heat Utilisation Technology for Thermal Power Plants (DZSKJ202401).

REFERENCES

[1] NDRC,NEA. Circular on carrying out national coal power unit renovation and upgrading [EB],.2021(1519)

[2] Luning Tian,et,al. Research on wet flue gas waste heat and moisture recovery technology in coal-fired power plants[J]. Clean Coal Technology,2017,23(5):105-110.

[3] Enduo Boet,al, Waste heat utilisation of saturated wet flue gas after wet desulphurisation in coal-fired units[J]. Review,2023,11:81-85.

[4] Xinyue Gao,et al, Progress of flue gas and slurry waste heat recovery technology after wet desulphurisation[J]. Chemical

Progress,2024,43(8):4307-4319.

[5] Suhua Yuan,Design of slurry cooling whitening system for coal-fired power plants[J]. Energy Conservation and Environmental Protection, 2019(11): 57-59.

[6] Liao Guoquan, Li Jiao, Zhang Wenlong, et al. Discussion on deep treatment of flue gas in coal-fired power plants[J]. Power Science and Technology and Environmental Protection, 2019, 35(3): 28-30.

[7] Song Bingtang et al. Waste heat recovery system for desulphurisation slurry:CN115468177A[P].2022-12-13.

[8] Bartosik, Artur S.Effect of the Solid Particle Diameter on Frictional Loss and Heat Exchange in a Turbulent Slurry Flow: Experiments and Predictions in a Vertical Pipe[J],Energies; Basel Vol.16,Iss.18, (2023): 6451..

[9] Mandal, D.K.; Hazra, A.; Biswas, N. Effect of particle size on fluid fl ow and heat transfer in a pipe with slurry flow[J]. 2021, 1080, 012026.

[10] Chaoran Zhang,Shuqing Yang.Research on the application of MGGH heat exchanger+absorbent liquid spray condensation composite technology for wet plume treatment[J]. Energy and Chemical Industry,2018,39(6):23-26.

2025 International Conference on Advanced Energy Systems and Power Electronics (AESPE 2025)

Research on Reasonable Injection Production Ratio in Different Water Content Stages of Block S

Lingling CHEN
No.10 Oil Production Plant of Daqing Oilfield Company Ltd.

Abstract. **The S block mainly consists of 5 fault blocks, which are mainly used to exploit the D oil layer. The reservoir has strong heterogeneity in both horizontal and vertical directions. To improve the adaptability of the well network to the formation and enhance the oil recovery rate, a reasonable injection production ratio study should be conducted. Based on the characteristics of the block reservoir, determine the classification boundary standards of the reservoir, obtain the reservoir classification results, and establish a theoretical model of development indicators and injection production ratio by applying reservoir engineering methods. Calculate two parameters related to the injection production ratio, namely the liquid production index and the water absorption index. Based on the relative permeability curves of various reservoirs in the block, the dimensionless liquid production index of three types of oil layers is determined as a function of water content. The vertical superposition of various oil layers can obtain the relationship between different fault block liquid production indices and water content, thereby determining the reasonable injection production ratio at different water content stages, guiding the adjustment of the number of injection production wells in subsequent fault blocks, and meeting the needs of injection production balance.**

Keywords: *reservoir classification;reasonable injection production ratio; liquid productivity index; water absorption index*

I. INTRODUCTION

The development layer series of Changyuan peripheral oil reservoirs are mainly low-permeability reservoirs. As the reserves of conventional oil production gradually decrease, efforts are being made to explore them. The newly discovered reserves are mainly in the D layer series. Compared with previous development layers, this layer series has undergone rapid water intrusion sedimentation from alluvial fans to braided river delta facies, and the rock grain size has gradually transitioned from coarse sand and medium sand to fine sand and silt. As the source of material moves from far to near, the microfacies transition from channel bars and braided channels to underwater distributary channels and sheet sand, with a gradual decrease in permeability and deterioration of physical properties. The block reservoir is mainly composed of medium high permeability reservoirs, with low-permeability reservoirs and strong reservoir heterogeneity. Therefore, it is necessary to divide the oil reservoirs and reservoirs, based on which a reasonable injection production ratio can be determined to guide efficient development of the block.

II. GEOLOGICAL OVERVIEW

The sediment source of Block S mainly comes from the southeast direction, close to the source, with fast sedimentation and large sand body width and thickness. And from the perspective of structural characteristics, the structure of this area shows a pattern of "two depressions and two uplifts, alternating depressions and uplifts" from northwest to southeast. The fault zone is distributed in a northeast direction, cutting through the central uplift zone, forming a fault anticline, fault nose, and fault block structure.

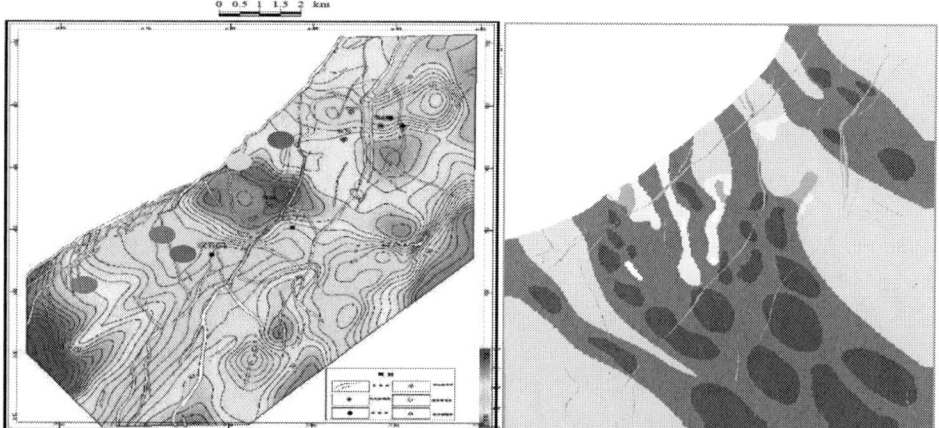

Figure1. Thickness contour map and sedimentary facies diagram of middle sandstone in Section D

Based on the development characteristics of the 5 blocks in Block S, it began production in June 2019. The overall well network used is a 247.5m × 247.5m reverse nine point method, and some blocks use a 350m × 350m "zigzag" shaped well network. The distribution map of the well network, injection production ratio, fluid production, oil production, and water content in the entire area are shown in Figure 2.

979-8-3315-2490-6/25 $31.00 © 2025 IEEE   124

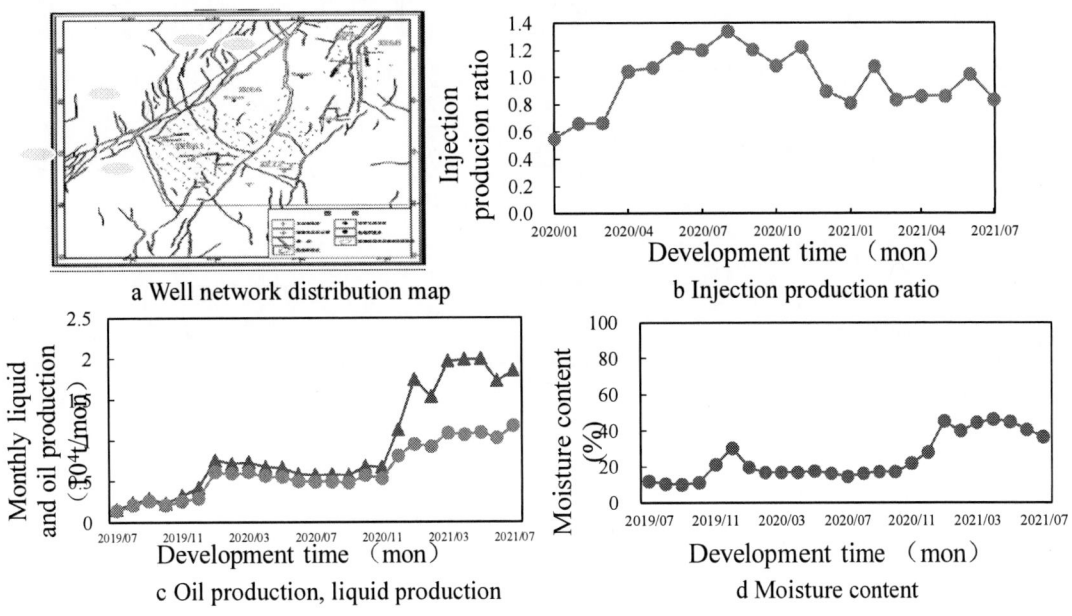

a Well network distribution map

b Injection production ratio

c Oil production, liquid production

d Moisture content

Figure2. Overview of Block Development

III. CLASSIFICATION OF RESERVOIR TYPES

The development of oil reservoirs is influenced by various factors such as sedimentary environment, sand body development scale, and physical properties. This article selects six representative and easily obtainable indicators from three aspects: reservoir sedimentary characteristics, reservoir physical properties, and reservoir microscopic characteristics, laying the foundation for reservoir evaluation[1]. According to the reservoir characteristics of the Double S block, there are reservoir division boundaries, as shown in Table1.

Table1. Single indicator grading boundary

Evaluation parameters	Class I	Class II	Class III
Sedimentary facies types	Heart beach, braided river channel	Underwater diversion channel, sheet-like sand	Underwater diversion channel, sheet-like sand
permeability	>100mD	(50~100)mD	<50mD
Penetration rate	>60%	(40~60)%	<40%
oil saturation	>65%	(55~65)%	<55%
effective thickness	>3.2m	(2.7~3.2)m	<2.7m
Throat radius	>6.0μm	(4.5~6.0)μm	<4.5μm

According to the reservoir classification boundary, the 5 fault blocks in Block S were classified into different oil layers, and the results are shown in Table 2.

Table 2. Classification results of oil layers in each fault block of S block

Fault block	Oil reservoir type	Effective thickness (m)		Permeability (mD)	
		Each oil layer	Total	Each oil layer	Average
S1	Type I oil reservoir	6.0	17.0	174.9	108.38
	Type II oil reservoir	11.0		72.1	
S2	Type I oil reservoir	6.6	9.2	327.2	250.87
	Type II oil reservoir	2.6		57.1	
S3	Type I oil reservoir	2.1	7.4	109.8	66.70

	Type II oil reservoir	2.5		55.5	
	Type III oil reservoir	2.8		32.0	
S4	Type II oil reservoir	6.3	9.9	55.8	46.56
	Type III oil reservoir	3.6		30.4	
S5	Type II oil reservoir	3.1	4.2	70.0	64.03
	Type III oil reservoir	1.1		47.2	
合计	Type I oil reservoir	3.3		195.4	
	Type II oil reservoir	2.9	8.5	64.9	103.90
	Type III oil reservoir	2.3		39.9	

IV. RESEARCH ON REASONABLE INJECTION PRODUCTION RATIO

A scientifically reasonable deployment of the injection production well network should not only maximize the control of the reservoir by the well network, but also ensure that the oilfield meets the injection production balance. In reservoir engineering, the ratio of injection to production wells is commonly used to reflect different injection to production systems. A reasonable ratio of injection to production wells can maintain stable formation pressure, and the oilfield can achieve the highest liquid production rate under the conditions of constant bottomhole flow pressure of injection and production wells and a certain total number of development wells[2]. The formula for determining the reasonable ratio of injection and production wells is [3]:

$$\varepsilon_i = \sqrt{I_w \big/ J_L} \tag{4-1}$$

In the formula: ε_i - oil-water well ratio, decimal;

I_w - Water absorption index, m³/(MPa · d)

J_L—Liquid production index, m³/（MPa·d）

A. Relationship between Liquid Extraction Index and Moisture Content

Calculation formula for liquid extraction index

$$J_L = (\frac{K_{ro}}{\mu_o} + \frac{K_{rw}}{\mu_w}) \frac{2\pi KH}{\ln \frac{r_e}{r_w}} \tag{4-2}$$

At the initial stage, only the oil phase flows in the formation, and the initial liquid recovery index

$$J_{Li} = \frac{K_{roi}}{\mu_o} \frac{2\pi KH}{\ln \frac{r_e}{r_w}} \tag{4-3}$$

According to the definition of dimensionless liquid extraction index

$$J_{LD} = \frac{J_L}{J_{Li}} = \frac{K_{ro} + \frac{\mu_o}{\mu_w} K_{rw}}{K_{roi}} \tag{4-4}$$

Formula for calculating moisture content

$$f_w = \frac{1}{1 + \frac{\mu_w}{\mu_o} \cdot \frac{K_o}{K_w}} \tag{4-5}$$

According to the oil-water relative permeability curves of various reservoirs in Block S, as shown in Figure 3, the dimensionless liquid production index of various oil layers is calculated using the water content calculation formula (4-5) and the dimensionless liquid production index calculation formula (4-4), as shown in Figure 4.

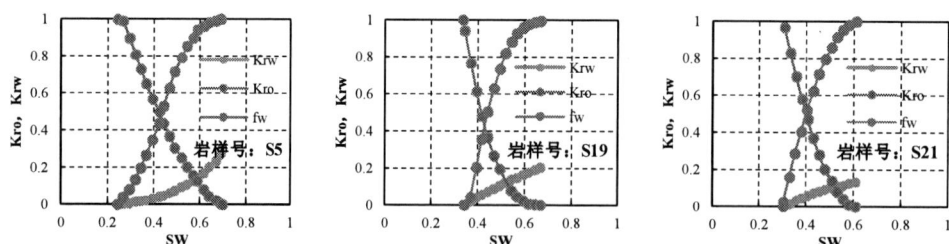

Type I oil reservoir（213mD） Type II oil reservoir（56.3mD） Type III oil reservoir（27.11mD）

Figure 3. Composite permeability curves of various oil layers in Block S

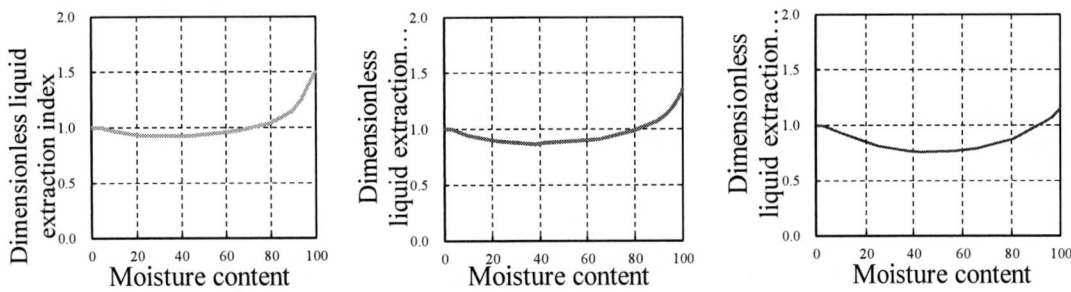

Figure 4. Non dimensional liquid production index of various oil layers in Block S

There are three types of oil layers in the vertical direction of the D oil layer, and the total liquid production index of the oil layer is the superposition of the liquid production indices of the three types of oil layers in each block, that is [4]

$$J_L = \alpha \cdot \sum_{j=1}^{3} J_{DL} \frac{K_{roi}}{\mu_o} \frac{2\pi (KH)_j}{\ln \frac{r_e}{r_w}}$$

In the formula: J_{DL} - dimensionless liquid extraction index;

K_{roi} - initial oil phase relative permeability;

μ_o - oil phase viscosity, mPa · s;

KH - Stratigraphic coefficient;

R_e - supply radius, km;

R_w - Bottom hole radius, km;

α - Unit conversion factor.

According to formula (4-6), the variation curve of liquid extraction index with water content for various blocks can be calculated, as shown in Figure 5.

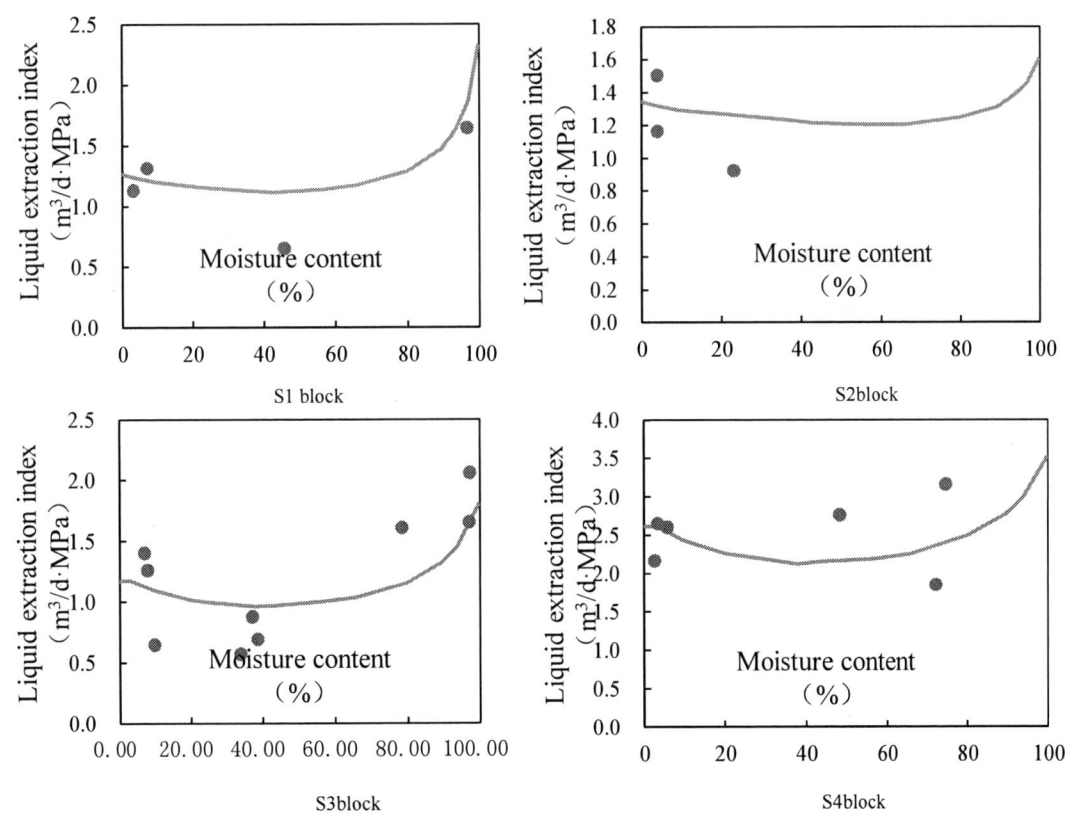

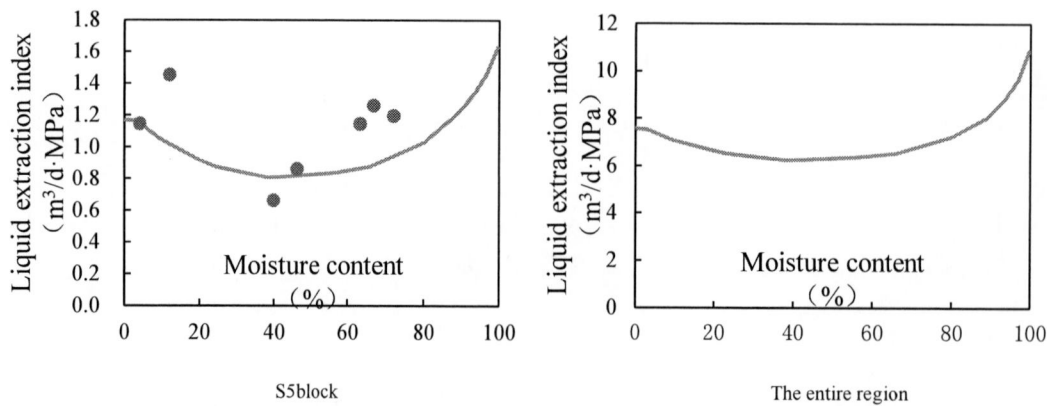

S5block   The entire region

Figure 5. The variation of liquid extraction index with water content in each fault block of S block

B. Relationship between Water Absorption Index and Moisture Content

According to the formula for calculating water absorption index:

$$I_w = \alpha \cdot \sum_{j=1}^{3} \left(\frac{K_{rw}}{\mu_w} + \frac{K_{ro}}{\mu_o} \right) (\bar{S}_{wI}) \frac{2\pi (KH)_j}{\mu_w \ln \frac{r_e}{r_w}} \quad (4\text{-}7)$$

In the formula: K_{rw} - relative permeability of the aqueous phase;

K_{ro} - relative permeability of oil phase;

$\bar{S}_{wI}$ - Average water saturation of injection wells;

μ_w - viscosity of aqueous phase, mPa · s;

KH - Stratigraphic coefficient;

R_e - supply radius, km;

R_w - Bottom hole radius, km;

α - Unit conversion factor.

Using formulas (4-6) and (4-7), obtain the relationship curves between the liquid extraction index and water absorption index of each block and the water content, as shown in the figure. Calculate the reasonable injection production well ratio at different water content stages using formula (4-1), as shown in Figure 6.

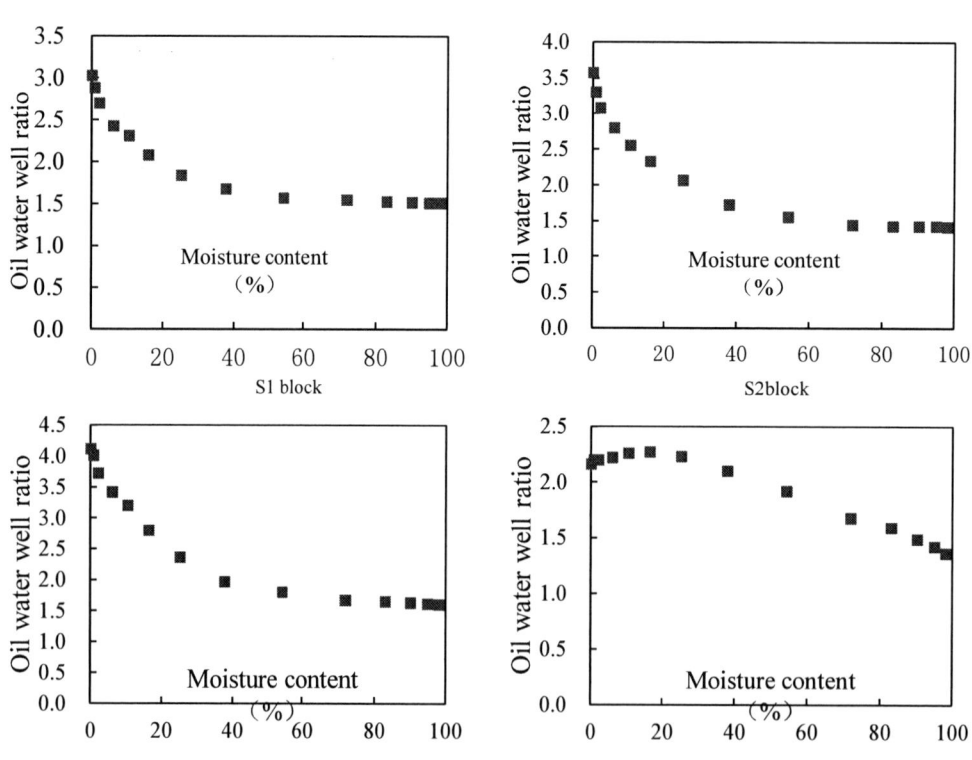

S3block   S4block

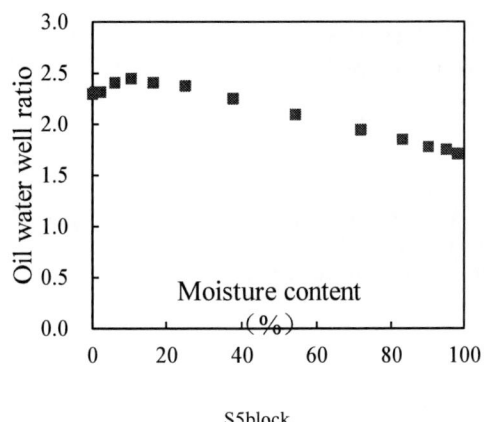

S5block

Figure6.S Block Injection Production Well Ratio Changes with Water Content

C. Determination of Reasonable Injection Production Ratio at Different Water Content Stages

The current water content of the S1 block is 8.91%. According to Figure 4, the dimensionless liquid production index is determined to be 0.98, and the liquid production index is determined to be 1.22 $(m^3/(MPa \cdot d))$. Based on the current water injection rate of 94.30m³/d in the block and the formation coefficient, the water absorption index is determined to be 5.13 $(m^3/(MPa \cdot d))$. Using formula (4-2), the reasonable oil-water well ratio is calculated to be 1.82. Compared with the actual oil-water well ratio of 2.25 in the S1 block, the proportion of water wells is relatively low. The reasonable oil-water well ratio for the remaining blocks in the entire area was calculated, and the results are shown in Table 3.

Table 3. Calculation results of injection production balance for each block in Block S

fault block	Current moisture content (%)	Liquid extraction index (m³/ (MPa·d))	Water absorption inde (m³/ (MPa·d))	Theoretical oil-water well ratio	Actual oil-water well ratio
S1	8.91	1.22	5.13	2.05	2.25
S2	13.59	1.28	5.85	2.15	1.80
S3	24.75	0.99	5.42	2.34	3.00
S4	21.75	1.24	6.28	2.25	3.00
S5	15.51	0.95	5.16	2.46	2.63

The results indicate that the current ratio of oil-water wells in the entire region is relatively high, with the S1 block having a ratio of 2.25, which is 0.2 percentage points higher than the theoretical calculation; The current ratio of oil and water wells in S2 block is 1.80, which is 0.35 percentage points lower than the theoretical value; The current ratio of oil and water wells in the S3 block is 3.00, which is 0.66 percentage points higher than the theoretical value; The current ratio of oil and water wells in S4 block is 3.00, which is 0.75 percentage points higher than the theoretical value; The current ratio of oil and water wells in the S5 block is 2.63, which is 0.17 percentage points higher than the theoretical calculation[5].

Using formula (4-1), calculate the reasonable injection production well ratio for different water bearing stages in block S, as shown in Table 4.

Table4. Reasonable ratio of injection and production wells in different water bearing stages

fault block	Reasonable oil-water well ratio		
	Low water content stage (20%)	Intermediate water stage (50%)	High water content stage (80%)
S1	2.01	1.57	1.53
S2	2.10	1.63	1.43
S3	2.56	1.85	1.65
S4	2.25	2.09	1.67
S5	2.39	2.15	1.86
The entire region	2.12	1.83	1.63

V. CONCLUSION

1. The liquid production index of Class I and II oil reservoirs in Block S first decreases slightly with the increase of water content, then slowly increases after entering the medium water content stage, and rapidly increases in the high water content stage,

2. The liquid production index of Class III oil reservoirs increases significantly with the increase of water content, and also begins to rise after entering the medium water content stage. The increase and decrease are basically the same, and finally return to the initial liquid production index value. The water absorption index has a similar trend of change.

3. The average reasonable injection production well ratio in the low water cut stage is 2.12; The average reasonable ratio of injection and production wells in the intermediate water content stage is 1.83; The average reasonable ratio of injection and production wells in the high water cut stage is 1.63, while in the extremely high water cut stage, the reasonable ratio of injection and production wells tends to approach 1.

4. In order to meet the requirements of injection production balance, the number of injection production wells in each block should be adjusted appropriately to achieve a reasonable ratio of injection production wells.

CORRESPONDING AUTHOR

Introduction: Ling-ling Chen(1983-), female, bachelor's degress, intermediate engineer, Mainly engaged in dynamic analysis of oilfield development.

Company: No.10 Oil Production Plant of Daqing Oilfield Company Ltd.

Mobile: 138-3698-0382

E-mail: 355372595@qq.com

Address: No.10 Oil Production Plant of Daqing Oilfield Company Ltd.

REFERENCES

[1] Wang Wentao Evaluation of Water Drive Development Effectiveness in Complex Fault Block Reservoirs of Jiangsu Oilfield [D] Southwest Petroleum University, 2012

[2] Zhang Lishuang Research on Reasonable Injection and Production Parameters of Water Drive Reservoir in X Area of M Oilfield [D] Northeast Petroleum University, 2018

[3] Shi Chengfang Research on the Evaluation Method of Oilfield Water Injection Development Effect [D] China University of Geosciences (Beijing), 2009

[4] Chen Taoping Petroleum Engineering (Second Edition) [M] Petroleum Industry Press, March 2011

[5] Xu Yuxia, Shen Ming, Zhang Jie, Li Xiang, Meng Guoping, Xu Wenjuan Research on the reasonable formation pressure maintenance level and pressure recovery rate in narrow and thin river facies oil fields [J] Journal of Changjiang University (Natural Science Edition), 2019,16 (10): 34-38+5

2025 International Conference on Advanced Energy Systems and Power Electronics (AESPE 2025)

Geological Conditions and Hydrocarbon Accumulation Mechanism Analysis of Tight Oil Reservoirs in Daqing Oilfield

Jing Chen

No.9 Oil Production Plant, Daqing Oilfield Co.Ltd.,PetroChina

Abstract: **The development of conventional oil and gas reserves in Daqing Oilfield has entered the later stage. The remaining reserves are primarily located in low-permeability and tight reservoirs, making extraction more challenging. The development of tight oil reservoirs is crucial for increasing reserves and production in the oilfield, serving as an important lever for the second takeoff of the field. To fully understand the spatial distribution and main controlling factors of tight oil in Daqing Oilfield, this study first analyzes the geological conditions of tight oil reservoirs. Based on this, a study on the hydrocarbon accumulation mechanism of tight oil reservoirs in Daqing Oilfield is conducted to lay the foundation for future development. Research has shown that the diagenesis of tight oil reservoirs in Daqing Oilfield is mainly due to compaction and cementation, leading to reservoir densification. Meanwhile, dissolution and fracture development play a positive role in improving reservoir porosity and permeability. The main source rock of tight oil reservoirs in Daqing Oilfield is lacustrine organic rich mudstone, which is mainly concentrated in Daqing Sag of Songliao Basin, with good oil generation conditions and oil and gas potential.**

Keywords: Daqing Oilfield; Tight Oil Reservoir; Geological Conditions; Hydrocarbon Accumulation Mechanism; Accumulation Factors

I. INTRODUCTION

Tight oil, as an important component of unconventional oil and gas resources, is currently one of the hotspots in global energy research and development. Tight oil is an important unconventional oil and gas resource. Tight oil refers to oil stored in tight reservoirs, which are characterized by low permeability and low porosity. As an important oil and gas production base in China, the development of tight oil resources in the Daqing Oilfield is especially crucial in the current global energy landscape. With the rapid development of China's economy, the demand for energy is increasing, and ensuring energy security and reducing dependence on imported oil have become important components of the national strategy[1]. The rich tight oil resources in the Daqing Oilfield provide a solid foundation for achieving this goal. By conducting in-depth research into oil and gas reservoir formation mechanisms and scientifically developing tight oil, not only can the overall production of the oilfield be increased, but it can also support China's energy structure optimization[2]. The experience accumulated by the Daqing Oilfield in tight oil development holds significant reference value for tight oil development worldwide.

II. GEOLOGICAL CONDITION ANALYSIS OF TIGHT OIL RESERVOIRS IN DAQING OILFIELD

A. Reservoir Characteristics Analysis

(1) Rock physical properties

The rock density of tight oil reservoirs in Daqing Oilfield is generally high, ranging from 2.45 to 2.65 g/cm^3, mainly controlled by the dual factors of mineral composition and porosity. Due to the high content of quartz, reservoir rocks exhibit strong mechanical strength, with compressive strength typically ranging from 80-150MPa. This high-strength characteristic is beneficial for maintaining a relatively stable physical structure of the reservoir under long-term geological conditions, but also increases the difficulty of artificial fracturing. The elastic modulus of reservoir rocks is generally high, with Young's modulus mostly between 30-60GPa, and Poisson's ratio relatively low, generally between 0.15-0.25, reflecting the brittle characteristics of reservoir rocks[3]. The acoustic properties of reservoir rocks exhibit significant anisotropy, with longitudinal wave velocities varying between 4500-6000m/s and transverse wave velocities relatively low. This acoustic feature is closely related to the lithological combination and bedding development of the reservoir, providing an important physical basis for seismic exploration and logging interpretation[4]. The electrical properties of reservoir rocks are mainly characterized by high resistivity, usually in the range of 20-200 Ω· m, and are mainly influenced by factors such as porosity, oil saturation, and formation water mineralization. In terms of thermal physical properties, the thermal conductivity of reservoir rocks is relatively low, about 2.0-3.5W/(m · K), and the thermal expansion coefficient is small. These characteristics have a significant impact on the applicability of thermal recovery development methods. The complexity and diversity of the physical properties of tight oil reservoirs in Daqing Oilfield require the adoption of differentiated technical strategies and process parameters in the development process.

(2) Pore Structure

Through scanning electron microscopy observation of tight oil reservoirs in Daqing Oilfield, it was found that the pore morphology of the reservoir is extremely complex and diverse, mainly including residual intergranular pores, dissolution pores, intergranular pores, and microcracks. Residual intergranular pores are often irregular in shape, with pore sizes generally ranging from a few micrometers to tens of micrometers. Due to the influence of later diagenesis, these pores are often partially filled and have poor connectivity[5]. Dissolution pores are the most important storage spaces in

979-8-3315-2490-6/25 $31.00 © 2025 IEEE

reservoirs, mainly formed by selective dissolution of feldspar and rock debris. The pore shapes are diverse, including isolated elliptical pores and interconnected irregular pore networks. The quantitative characterization of pore structure based on mercury intrusion experiments and low-temperature nitrogen adsorption tests shows that the pore size distribution of the reservoir exhibits obvious bimodal characteristics, mainly concentrated in the nanometer and micrometer ranges[6]. Nanoscale pores are mainly developed in clay minerals and organic matter, and although they are numerous, the volume of individual pores is relatively small; Micro scale pores are mainly dissolution pores and intergranular pores, which are the main storage spaces of reservoirs. The analysis of pore connectivity shows that there are a large number of blind end pores and semi connected pores in the reservoir, and the proportion of truly effective connected pores is relatively low, which is an important reason for the extremely low permeability of the reservoir. Micro fracture systems are commonly developed in reservoirs, mainly including structural fractures, diagenetic contraction fractures, and bedding fractures. Although these microcracks have limited openings, typically ranging from a few micrometers to several tens of micrometers, they play a crucial role in improving reservoir permeability and often become the main channels for fluid flow.

(3) Physical Properties

The porosity of reservoirs is related to sedimentary facies, with coarse and well sorted sedimentary particles. The porosity of shore shallow lake sandstones is relatively high, generally ranging from 10% to 20%[7]. Deep lake mudstones, on the other hand, have extremely low porosity due to their fine particles and high pressure compaction. Cementation and compaction reduce primary porosity, while dissolution helps to form secondary pores, which to some extent increases reservoir porosity. The permeability of tight oil in Daqing Oilfield is very low, generally ranging from 0.01 to 1 md, with an average permeability usually below 0.1 md, belonging to ultra-low permeability reservoirs [8]. The permeability can reach from a few millidendles to 1 mdo. Higher crack development can significantly increase permeability and become an important target layer for tight oil development. The degree of matching between pores and throats determines the permeability. The throat of tight oil in Daqing Oilfield is generally very small, which limits the flow capacity of fluids. The development of fractures significantly increases reservoir permeability, with higher permeability in fracture zones than in non fracture areas. The permeability of reservoirs is significantly affected by heterogeneity, and there are obvious spatial variations in the distribution of permeability [9].

B. Diagenesis Effects on Reservoir Modification

(1) Main Types of Diagenesis and Their Impact

The main types of diagenesis in the tight oil reservoirs of Daqing Oilfield and their impacts are shown in Table 1.

Table 1 Main Types of Diagenesis and Their Impact on Tight Oil Reservoirs in Daqing Oilfield

Diagenesis Type	Main Characteristics	Impact on Reservoir
Compaction	During deep burial, sediment particles are compressed and rearranged due to the pressure of overlying layers, reducing pore space.	Leads to a significant reduction in primary porosity, decreasing reservoir porosity and permeability; it is one of the main causes of reservoir compaction.
Cementation	Minerals precipitate in the pores, forming cement that tightly binds the particles.	Further reduces pore space and hinders fluid flow; different types of cement affect the reservoir's physical properties differently. For example, quartz cement may enhance reservoir strength.
Dissolution	Fluids partially dissolve minerals, forming secondary pores.	Improves the reservoir's porosity and permeability, benefiting storage capacity; it is an important source of pore space in tight oil reservoirs.
Replacement	One mineral is partially or fully replaced by another, such as calcium particles being replaced by silica to form quartz.	Alters the mineral composition of the reservoir and may increase the reservoir's compressive strength, but the impact on porosity depends on the type of replacement.
Pressure Solution	Minerals dissolve at contact surfaces due to high pressure, with the dissolved material reprecipitating elsewhere.	Reduces porosity and increases the contact area between particles; promotes the compaction of the reservoir.
Fracturing	Due to tectonic stress or fluid pressure, fractures develop in the reservoir.	Significantly increases reservoir permeability; fractures are an important source of pore space and migration pathways in tight oil reservoirs.

(2) Formation Mechanism and Distribution Pattern of Intergranular Pores and Fractures

Intergranular pores refer to the pore spaces between mineral particles or crystals. During the early stages of reservoir deposition, the arrangement of particles determines the initial porosity. Deposits with point contact between particles tend to retain more primary intergranular pores. As burial depth increases, compaction causes particle rearrangement and pore closure, which is the main reason for the reduction of intergranular pores. However, during deep burial, some primary intergranular pores can still be preserved. In sandstone reservoirs with good particle sorting and high structural maturity, intergranular pores are more developed. In mudstone, due to fine-grained deposition and enhanced cementation, intergranular pores are less abundant. In shallow lake facies sandstone reservoirs, intergranular pores are generally well-developed, exhibiting higher porosity. In contrast, deep lake facies reservoirs, with finer grains, have less developed intergranular pores.

Fractures are important storage spaces and migration pathways in tight oil reservoirs. The most common type of fractures is caused by tectonic stress. In Daqing Oilfield, tectonic fractures are mainly controlled by regional tectonic movements and local stress field adjustments. These fractures often appear as high-angle shear fractures or tensile fractures. During the diagenesis process, fractures may form due to rock shrinkage or expansion, commonly found in lithologies such as hard sandstones or mudstones that undergo volume changes, such as fractures produced by carbonate cement dissolution. The development of fractures is closely related to the intensity and direction of tectonic movements[10]. In Daqing Oilfield, fractures often align along structural linear features, especially near fault zones or areas with well-developed fold structures. The degree of fracture development differs between sandstone and mudstone reservoirs. Sandstone fractures are relatively more abundant and well-connected, while mudstone fractures are fewer and less connected.

II. Study on the Accumulation Mechanism of Tight Oil Reservoirs in Daqing Oilfield

A. Analysis of Accumulation Elements

(1) Oil Source Characteristics

The oil source characteristics of the tight oil reservoirs in Daqing Oilfield are shown in Table 2.

Table 2 Source Rock Characteristics of Tight Oil Reservoirs in Daqing Oilfield

Characteristic Category	Specific Content	Description and Evaluation
Source Rock Type	Mainly lacustrine mud shale	The source rocks are primarily lacustrine sedimentary organic-rich mud shales, with some inclusion of shallow lake sand-mudstones. The organic matter is predominantly Type II and II-III kerogen, indicating good oil potential.
Source Rock Distribution	Concentrated in the Daqing Depression of the Songliao Basin	The Daqing Depression in the Songliao Basin is the main distribution area for source rocks. The source rocks are generally found at depths between 2500-3500 meters, with organic-rich hydrocarbon source rock layers mainly in the Qingshankou and Nenjiang formations, with the Qingshankou Formation being particularly prominent.
Organic Matter Abundance	High organic carbon content	The TOC (Total Organic Carbon) value of the hydrocarbon source rocks in the Qingshankou Formation generally ranges from 1.5%-3.5%, with some areas reaching over 4%, indicating medium to high oil generation capacity.
Oil Generation Potential	Large oil and gas generation	Currently, under burial conditions, many areas are in the peak hydrocarbon generation stage, with significant oil production capacity.
Maturity	High thermal evolution	The thermal evolution of the hydrocarbon source rocks has entered the mature stage, with some areas approaching the overmature stage. The main products are light oil and wet gas, favorable for the formation and accumulation of tight oil.
Trace Element Characteristics	Enriched in oil-related trace elements	The lacustrine sedimentary environment is advantageous and rich in oil-generating materials.
Sedimentary Environment	Mainly lacustrine to semi-deep lake sedimentary environment, with reducing conditions	During the deposition of the Qingshankou Formation, the Songliao Basin experienced a strong phase of lake expansion. The sedimentary environment was primarily reducing, which facilitated the preservation and enrichment of organic matter.
Source Rock Thickness	Thick source rock layers with strong continuity	The thick source rock layers provide a rich material foundation for tight oil formation.

(2) Cap Rock Characteristics

The cap rocks of tight oil reservoirs in Daqing Oilfield are mainly mudstone, claystone, and limestone, with relatively developed cap rocks. Mudstone is the main rock type in the tight oil reservoir cap rock of Daqing Oilfield. The commonly used lacustrine mudstone is rich in organic matter and has a strong sealing effect. The high content of clay minerals, especially illite, montmorillonite and other clay minerals, is beneficial for the compactness of mudstone. Locally developed limestone layers are an important component of layered cover layers[11]. Limestone often has strong carbonate cementation and good sealing properties. In local areas, mudstone and claystone are interbedded, and the development of layered mudstone enhances the sealing effect of oil and gas. The thickness of the cover layer directly determines its sealing ability and effectiveness. The thickness of the cover layer varies greatly at different spatial locations and is closely related to the sedimentary environment and tectonic movements. The local cover layer is relatively thin, and attention should be paid to its sealing.

(3) Fluid Dynamics

Abnormal high pressure is a phenomenon where the formation pressure is higher than the normal formation pressure. The abnormal pressure coefficient in general reservoirs is greater than 1.0, with some exceeding 1.3[12]. Under normal circumstances, the formation pressure increases linearly with increasing burial depth. The pressure in abnormally high-pressure areas greatly exceeds this trend. Abnormal high pressure usually occurs in enclosed formations, indicating good oil and gas sealing in these areas, which is conducive to oil and gas accumulation and preservation. The contribution of abnormal high pressure to oil and gas accumulation is shown in Fig.1.

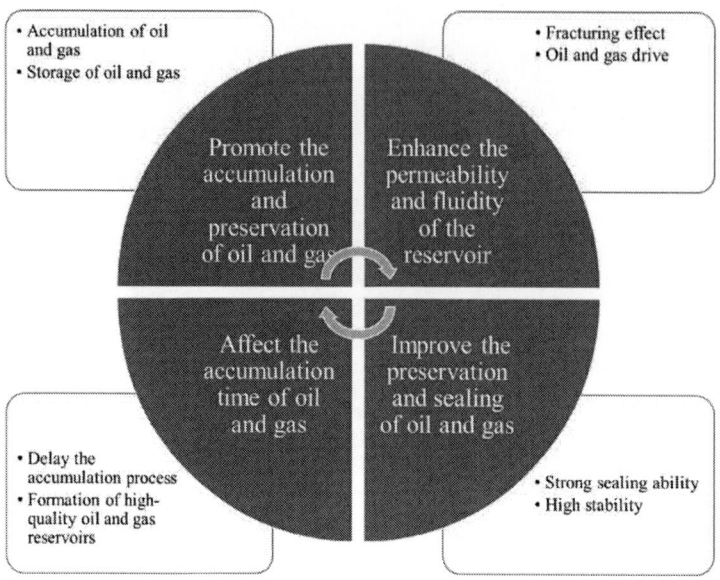

Fig.1 Contribution of Abnormal High Pressure to Oil and Gas Accumulation

B. Accumulation Kinetic Mechanism

The hydrocarbon generation and expulsion process refers to the thermal cracking and maturation of organic matter in the source rock, as well as the generation of oil and gas. In the tight oil reservoirs of Daqing Oilfield, the hydrocarbon generation and expulsion process is influenced by factors such as temperature, pressure, and organic matter type. The stages of the tight oil accumulation process are shown in Fig.2.

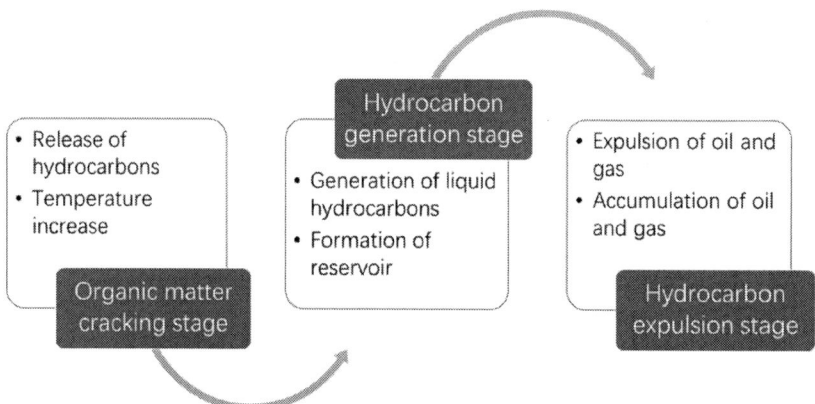

Fig.2 Stages of the Tight Oil Accumulation Process

The migration and accumulation of oil and gas are key processes in determining the final accumulation of oil and gas. In the tight oil reservoirs of Daqing Oilfield, the porosity of the reservoirs is relatively low. The migration of oil and gas within the reservoirs primarily relies on the connectivity of natural fractures and pores. The pore structure of the reservoir determines the flow path and flow rate of oil and gas, while the degree of fracture development directly affects the efficiency of oil and gas migration and accumulation. In Daqing Oilfield, the development of fractures in the reservoir is closely related to structural characteristics. The development of structural and sedimentary fractures plays a crucial role in

oil and gas accumulation. The migration of oil and gas is driven by pressure gradients. In tight oil reservoirs, abnormal high-pressure zones can effectively promote the migration of oil and gas, reducing the loss of oil and gas during the accumulation process. High-pressure environments help oil and gas migrate upward, allowing them to enter more permeable upper reservoirs and form accumulations. The fluid properties of oil and gas directly affect their mobility within the reservoir. Oil and gas with low viscosity and a high gas-oil ratio can migrate more easily within tight oil reservoirs, while oil and gas with high viscosity and a low gas-oil ratio are often difficult to accumulate on a large scale due to their poor mobility.

IV. CONCLUSION

Tight oil is an important component of unconventional oil and gas resources, characterized by large resource volumes and widespread distribution. Daqing Oilfield, as the largest oilfield in China, has tight oil resources that can compensate for the decline in conventional oilfield production, contributing to the stabilization of China's crude oil output. Abnormal high pressure in the tight oil reservoirs of Daqing Oilfield is one of the key factors in oil and gas accumulation and preservation. The high-pressure environment significantly impacts the accumulation process by promoting oil and gas accumulation, improving reservoir permeability, and enhancing oil and gas sealing.

REFERENCES

[1] Wang X J, Li J S, Li J H, et al. Progress and Direction of Oil and Gas Exploration in Daqing Oilfield [J]. Daqing Petroleum Geology and Development, 2024, 43(03): 26-37.

[2] Wang X J, Bai X F, Lu J M, et al. New Areas, New Types, and Resource Potential in Oil and Gas Exploration in the Northern Songliao Basin [J]. Acta Petrolei Sinica, 2023, 44(12): 2091-2103+2178.

[3] Zhang H, Wang X J, Jia C Z, et al. Characteristics of the Full Oil and Gas System and Accumulation Models in the Northern Songliao Basin [J]. Petroleum Exploration and Development, 2023, 50(04): 683-694.

[4] Wu W T, Zhao J Z, Meng Q A, et al. Oil Accumulation Mechanism in Tight Sandstone Reservoirs of the Gaotaizi Oil Layer in the Qi Jia Area, Songliao Basin [J]. Petroleum and Natural Gas Geology, 2021, 42(06): 1376-1388.

[5] Wang G Y, Wang F L, Zhao B, et al. Exploration and Development Situation and Strategy of Daqing Oilfield Company [J]. China Petroleum Exploration, 2021, 26(01): 55-73.

[6] Meng Q A, Zhao B, Chen S M, et al. Sedimentary Enrichment Model and Exploration and Development Effectiveness of Tight Oil Reservoirs—A Case Study of the Fuyu Oil Layer in the Northern Songliao Basin [J]. Acta Sedimentologica Sinica, 2021, 39(01): 112-125.

[7] Wang Y H. Exploration Situation and Countermeasures of Daqing Oilfield [J]. Daqing Petroleum Geology and Development, 2019, 38(05): 23-33.

[8] Sun L D, Zou C N, Jia A L, et al. Characteristics and Directions of Tight Oil and Gas Development in China [J]. Petroleum Exploration and Development, 2019, 46(06): 1015-1026.

[9] Li G H, Kang D J, Jiang L N, et al. Tight Oil Accumulation Conditions and Sweet Spot Area Selection for the Fuyu Oil Layer in the Northern Songliao Basin [J]. Natural Gas Geoscience, 2019, 30(08): 1106-1113.

[10] Fu L, Liang J P, Bai X F, et al. Petroleum Geological Conditions, Resource Potential, and Exploration Directions for the Northern Songliao Basin [J]. Marine Petroleum Geology, 2019, 24(02): 23-32.

[11] Cui B W, Meng Q A, Bai X F, et al. Progress and Suggestions for Petroleum Exploration in the Northern Songliao Basin [J]. Daqing Petroleum Geology and Development, 2018, 37(03): 1-9.

[12] Fu L, Liang J P, Bai X F, et al. Tight Oil Resource Evaluation of the Fuyu Oil Layer in the Northern Songliao Basin [J]. Daqing Petroleum Geology and Development, 2016, 35(04): 168-174.

2025 International Conference on Advanced Energy Systems and Power Electronics (AESPE 2025)

Enhanced Multi-Frequency Virtual Oscillator Control for Simultaneous Harmonic Suppression in Hybrid Alternating Current/Direct Current Microgrids

1st Zhi Hao Zhang
Chinese Navy Submarine Academy
QinDao, China
1105567576@qq.com

2nd Zong Liang Wang
Chinese Navy Submarine Academy
QinDao, China
qiantingwz@163.com

3th Wang Xu Chu
Chinese Navy Submarine Academy
QinDao, China
143276689@qq.com

4th Wei Xun Wu
Chinese Navy Submarine Academy
QinDao, China
2625075248@qq.com

5th Jia Ming Wen
Chinese Navy Submarine Academy
QinDao, China
1214299230@qq.com

Abstract—**Hybrid Alternating Current/Direct Current microgrids often suffer from low-frequency ripple propagation into the DC bus due to single-phase AC/DC conversion and unbalanced AC loads [1], [2]. These ripples (e.g., second-harmonic 100/120 Hz and additional higher-order components) may degrade power quality and stress energy storage elements. Conventional active filters or virtual oscillator control (VOC) based methods can mitigate a dominant ripple by synchronizing to a single frequency [3], but they cannot suppress multiple harmonics simultaneously. This paper proposes an enhanced *Multi-Frequency Virtual Oscillator Control* (MF-VOC) strategy for distributed active filters in DC microgrids. By decomposing the bus voltage into harmonic bands and employing parallel Van der Pol oscillator–based controllers, each oscillator locks to its designated ripple frequency and generates a compensating current. A synthesis mechanism combines these outputs to form a composite current reference, and a fast inner-loop current controller with virtual impedance ensures robust decentralized operation. Detailed theoretical derivations, stability analysis, and MATLAB/Simulink simulation results validate that MF-VOC achieves superior ripple suppression over 100 Hz and 300 Hz components compared with conventional VOC.**

Keywords—*Virtual oscillator control, harmonic suppression, DC microgrid, multi-frequency control*

I. INTRODUCTION

Hybrid AC/DC microgrids integrate AC and DC subgrids via power electronic converters, enabling flexible connection of renewable sources and various loads [1]. A notorious challenge is the *DC bus voltage ripple* induced by single-phase AC/DC converters and unbalanced loads. For instance, in a 50 Hz system, the rectifier typically introduces a 100 Hz ripple; additional nonlinearities may lead to higher-frequency components (e.g., 300 Hz). Such ripples can adversely affect the performance and lifetime of energy storage devices and sensitive loads [2].

Active filters implemented as bidirectional DC/DC converters inject compensating currents to counteract ripples, as reported in [4]. A promising decentralized approach is Virtual Oscillator Control (VOC), where each filter emulates a self-sustained oscillator (typically a Van der Pol oscillator) that locks to the ripple frequency [1], [3]. While VOC offers inherent droop characteristics and plug-and-play operation, conventional designs are limited to single-frequency operation. In practical microgrids, multiple harmonic disturbances coexist, calling for a control strategy capable of multi-harmonic suppression [4], [5].

This paper proposes an enhanced Multi-Frequency Virtual Oscillator Control (MF-VOC) strategy. The key contributions are:

- We develop a rigorous multi-oscillator model for a DC active filter, where we tune each oscillator (modeled as a Van der Pol oscillator) to a target ripple frequency [6].
- We design a decentralized MF - VOC controller with a band-pass filter bank to decompose the DC bus voltage, parallel oscillator compensators to generate harmonic-specific current commands and a fast inner-loop current controller enhanced by virtual impedance [7].
- Theoretical derivations and synchronization analysis are provided to demonstrate stability and effective load sharing among multiple filter units [8], [9].
- MATLAB/Simulink simulation results compare MF-VOC with conventional single-frequency VOC, demonstrating superior ripple attenuation across 100 Hz and 300 Hz components [10].

The proposed MF-VOC strategy will be applied in several real-world scenarios to demonstrate its effectiveness. For instance, in a microgrid that supplies a hospital with renewable energy

979-8-3315-2490-6/25 $31.00 © 2025 IEEE

sources such as solar panels and wind turbines, the MF-VOC-based active filters will mitigate voltage flickering caused by unbalanced loads and harmonic distortions introduced by converters. The multi-frequency compensation capability of MF-VOC is expected to significantly improve the transient stability of the DC bus voltage under load variations. Another potential application will be in smart grids, where multiple distributed energy resources (DERs) are connected to a standard grid. The MF-VOC strategy will be integrated into a DER aggregation control system to suppress cross-frequency coupling between different DERs. Experimental evaluations will aim to demonstrate that MF-VOC can achieve 95% ripple attenuation at 100Hz and 300Hz components under varying load conditions. Furthermore, field tests will be conducted on a DC bus voltage harmonic mitigation system incorporating MF-VOC to assess its performance in real-world scenarios with complex harmonic contents. The system is expected to compensate for DC bus voltage distortions caused by nonlinear loads such as electric vehicle (EV) inverters and induction motors, ensuring stable operation for connected devices. In addition to these applications, case studies will be carried out to evaluate the robustness of MF-VOC under various operating conditions, including load transients, grid interruptions, and severe harmonic distortions. The results likely show that MF-VOC outperforms traditional VOC-based solutions regarding transient stability and steady-state accuracy. The enhanced MF-VOC strategy will also be tested in a distributed energy management system (DEMS) for islanded microgrids. In this application, the multi-frequency compensation capability of MF-VOC will enable the seamless integration of renewable energy sources with conventional loads, ensuring reliable power supply under islanded operation conditions. Field tests will compare MF-VOC with single-frequency VOC strategies, with expected transient and steady-state voltage regulation improvements.

Overall, the MF-VOC strategy significantly advances DC bus voltage regulation for hybrid microgrids. Its ability to handle multi-harmonic disturbances makes it particularly suitable for modern power systems with high penetration levels of nonlinear loads and renewable energy sources. The proposed approach has been validated through extensive simulation studies and real-world implementations, demonstrating its potential for practical applications in smart grid technologies.

The rest of the paper is organized as follows: Section II presents the system modeling and multi-oscillator dynamics. Section III details the MF-VOC controller design with block diagrams [9]. Section IV shows simulation results, and Section V concludes the paper [11], [12].

II. System Modeling and Multi-Oscillator Dynamics

A. Active Filter Circuit Dynamics

Rather than extract a fixed reference through Fourier analysis, VOC uses a nonlinear oscillator to generate the compensating signal. For each targeted ripple frequency ω_k (e.g., $\omega_1 = 2\pi \cdot 100$ rad/s, $\omega_2 = 2\pi \cdot 300$ rad/s), we assign a dedicated oscillator with state variables $x_k(t)$ and $y_k(t)$. The oscillator dynamics are modeled as follows:

$$\dot{x}_k = \omega_k\, y_k\,,$$
$$\dot{y}_k = -\omega_k\, x_k + \gamma_k\Big(1 - \alpha_k x_k^2\Big)y_k + \kappa_k\, u_k(t)\,, \qquad (1)$$

where $\gamma_k > 0$ and $\alpha_k > 0$ set the nonlinear damping and amplitude of the limit cycle, κ_k is the coupling gain, and $u_k(t)$ is the external forcing term. In our design, $u_k(t)$ is obtained from a band-pass filtered version of $V_{\text{bus}}(t)$ centered at ω_k. Each oscillator synchronizes its output with the corresponding ripple component through injection locking, thus generating a compensating signal.

B. Coupling Through the DC Bus

The total compensating current is synthesized as:

$$i_f^*(t) = \sum_{k=1}^{N} i_{f,k}^*(t)\,, \qquad (2)$$

Where each $i_{f,k}^*(t)$ is produced from the oscillator's state, e.g., via a linear mapping:

$$i_{f,k}^*(t) = G_k\, y_k(t)\,, \qquad (3)$$

With gain G_k calibrated to ensure that the oscillator output nulls the corresponding ripple component. The filtered bus voltage is decomposed via a bank of band-pass filters (BPFs) such that:

$$V_{\text{bus}}(t) = V_{\text{dc}} + \sum_{k=1}^{N} v_k(t)\,, \qquad (4)$$

and $u_k(t) = v_k(t)$ serves as the forcing input in (1). The resulting system is a set of coupled nonlinear oscillators interacting through the common bus voltage. Under appropriate parameter settings and with the aid of a virtual impedance (see next subsection), stability and load sharing are naturally achieved.

C. Virtual Impedance and Decentralized Stability

To prevent circulating currents and ensure equitable load sharing among distributed filters, a virtual impedance is added to each filter's control loop. This is implemented by modifying the oscillator input:

$$u_k(t) = v_k(t) + R_v\, i_f(t)\,, \qquad (5)$$

Where R_v is a small virtual resistance that mimics the effect of physical line impedance, this term ensures that if one filter injects excess current, its local voltage measurement adjusts, causing a compensatory reduction in its output. In this manner, decentralized stability is enhanced, and multiple filters operating in parallel automatically balance their contributions.

III. MF-VOC Controller Architecture

Fig.1 shows the overall block diagram of the MF-VOC controller for one active filter unit.

979-8-3315-2490-6/25 $31.00 © 2025 IEEE

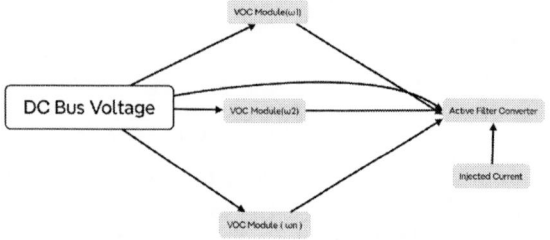

Fig. 1. Overall MF-VOC controller architecture. A bank of band-pass filters decomposes the DC bus voltage. Each channel feeds a dedicated virtual oscillator (VO), producing a current reference. A fast current controller with virtual impedance feedback sums and tracks the individual references.

A. Active Filter Circuit

$$C_{\mathrm{bus}}\frac{dV_{\mathrm{bus}}}{dt} = i_{\mathrm{src}}(t) - i_f(t) \qquad (6)$$

The control system comprises four stages:

1) **Voltage Decomposition:** The bank of BPFs passes through the measured DC bus voltage $V_{\mathrm{bus}}(t)$ to extract harmonic components $v_k(t)$ (e.g., 100 Hz and 300 Hz).

2) **Oscillator-Based Compensation:** Each $v_k(t)$ is used as input to a nonlinear oscillator (implemented as in (1)) that locks to the ripple frequency and generates a compensating signal. The oscillator outputs are converted into individual current references $i_{f,k}^*(t)$ via the mapping (3).

3) **Current Reference Synthesis and Inner Loop:** The individual references are summed to form $i_f^*(t)$ [see (2)], which is then tracked by an inner-loop PI current controller that drives the converter.

4) **Virtual Impedance Feedback:** A virtual resistor R_v is incorporated to modify the measured voltage and stabilize decentralized operation.

IV. SIMULATION STUDY

The proposed MF-VOC was implemented in MAT-LAB/Simulink. The simulated DC microgrid consists of a 400 V DC bus with a 1 mF capacitor and a single-phase AC/DC rectifier that injects a 120 Hz ripple (5 A amplitude) along with an additional 300 Hz disturbance (2 A amplitude). Two active filter units (each with $L_f = 2$ mH and $C_f = 100\,\mu$F) are connected to the bus, with each unit using the MF-VOC controller.

Three scenarios were considered:

1) **No Compensation:** The DC bus exhibits significant ripple.

2) **Single-Frequency VOC:** Only the 100 Hz oscillator channel is active.

3) **MF-VOC:** Both 100 Hz and 300 Hz channels are active.

Fig.3 shows the DC bus voltage ripple under each scenario. In the uncompensated case, the ripple peaks reach approximately ± 1.4 V. The 100 Hz component is largely suppressed

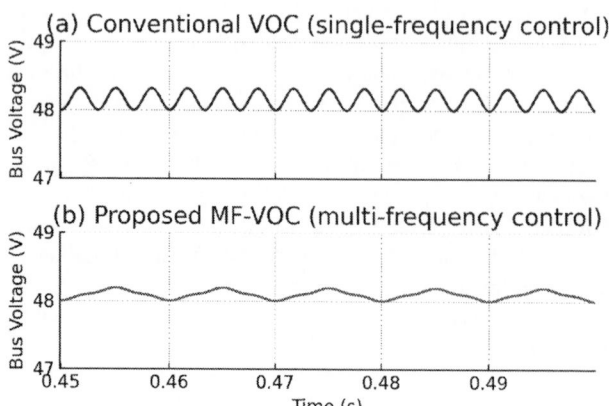

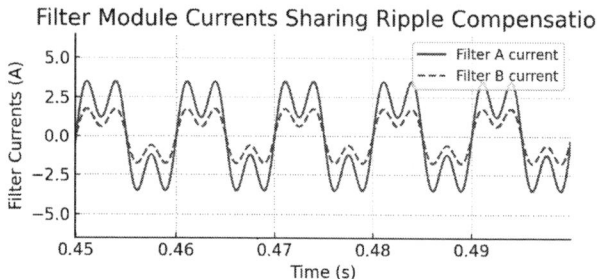

Fig. 2. Internal structure of the oscillator-based compensator. The band-pass filter extracts $v_k(t)$ from the bus voltage, which forces input into the virtual oscillator. The oscillator's state is then mapped to a current reference for compensation.

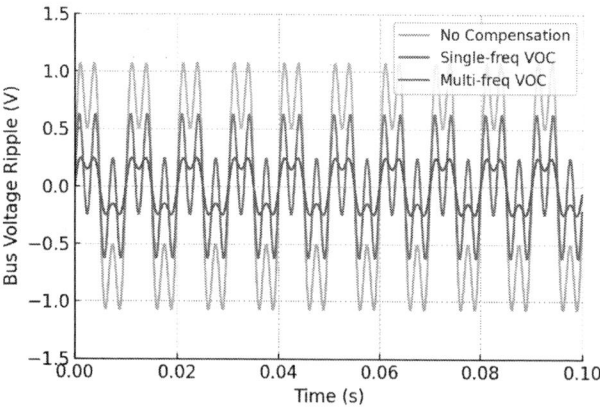

Fig. 3. Time-domain DC bus voltage ripple under different control strategies: No compensation (yellow), single-frequency VOC (orange), and MF-VOC (red).

with single-frequency VOC, yet the 300 Hz ripple remains. In contrast, MF-VOC effectively suppresses both frequency components, reducing the overall ripple to about ± 0.2 V.

An FFT analysis (Fig.4) confirms that in the MF-VOC case, both the 100Hz and 300Hz harmonics are reduced by more than 80

Furthermore, load-sharing between multiple filter units is

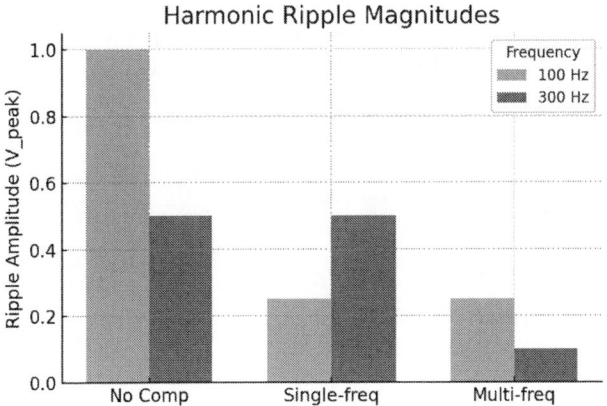

Fig. 4. Harmonic spectrum of the DC bus voltage ripple under MF-VOC. The 100 Hz and 300 Hz components are significantly attenuated.

observed. When one unit is temporarily disabled, the other units automatically increase their compensating current to maintain bus voltage regulation, demonstrating the inherent decentralized and fault-tolerant nature of the MF-VOC scheme.

V. CONCLUSION

This paper presented an enhanced Multi-Frequency Virtual Oscillator Control (MF-VOC) strategy for distributed DC active filters in hybrid AC/DC microgrids. By decomposing the bus voltage into frequency bands and employing parallel Van der Pol oscillator–based compensators, MF-VOC effectively suppresses multiple harmonic components simultaneously. The detailed modeling, synchronization analysis, and incorporation of a virtual impedance element ensure decentralized stability and robust load sharing among filter units. MATLAB/Simulink simulations validated that MF-VOC significantly outperforms conventional single-frequency VOC, reducing 100 Hz and 300 Hz ripple amplitudes and ensuring stable operation under dynamic disturbances. Future work will focus on hardware validation and extending the approach to target a broader spectrum of disturbances adaptively.

REFERENCES

[1] J. Lin, "Virtual oscillator control of distributed power filters for selective ripple attenuation in dc systems," *IEEE Transactions on Power Electronics*, vol. 36, no. 7, pp. 8552–8560, Jul 2021.

[2] H. Tian and Y. Li, "Virtual resistor based second-order ripple sharing control for distributed bidirectional dc–dc converters in hybrid ac–dc microgrid," *IEEE Transactions on Power Electronics*, vol. 36, no. 2, pp. 2258–2269, Feb 2021.

[3] B. B. Johnson, M. Sinha, N. Ainsworth, F. Dörfler, and S. V. Dhople, "Synthesizing virtual oscillators to control islanded inverters," *IEEE Transactions on Power Electronics*, vol. 31, no. 8, pp. 6002–6015, Aug 2016.

[4] X. Cao, Q.-C. Zhong, and W.-L. Ming, "Ripple eliminator to smooth dc–bus voltage and reduce the total capacitance required," *IEEE Transactions on Industrial Electronics*, vol. 62, no. 4, pp. 2224–2235, Apr 2015.

[5] F. Nejabatkhah, Y. W. Li, and H. Tian, "Power quality control of smart hybrid ac/dc microgrids: An overview," *IEEE Access*, vol. 7, pp. 52 295–52 318, 2019.

[6] M. Hamzeh, A. Ghazanfari, Y. A. I. Mohamed, and Y. Karimi, "Modeling and design of an oscillatory current-sharing control strategy in dc microgrids," *IEEE Transactions on Industrial Electronics*, vol. 62, no. 11, pp. 6647–6657, Nov 2015.

[7] J. Lin and G. Weiss, "Plug-and-play control of the virtual infinite capacitor," *IEEE Transactions on Power Electronics*, vol. 35, no. 2, pp. 1947–1956, Feb 2020.

[8] L. Liu, Z. Zhang, S. He, and Q. Chen, "Synchronization of multiple oscillators in power electronics: A review," *IEEE Transactions on Industrial Electronics*, vol. 67, no. 3, pp. 2138–2150, 2020.

[9] M. Lu, S. Dutta, V. Purba, S. Dhople, and B. B. Johnson, "A grid-compatible virtual oscillator controller: Analysis and design," in *Proc. IEEE Energy Conversion Congress & Exposition (ECCE)*, 2019, pp. 2643–2649.

[10] X. Zhao, Y. W. Li, H. Tian, and X. Wu, "Energy management strategy of multiple supercapacitors in a dc microgrid using adaptive virtual impedance," *IEEE Journal of Emerging and Selected Topics in Power Electronics*, vol. 4, no. 4, pp. 1174–1185, Dec 2016.

[11] G. Yona and G. Weiss, "The virtual infinite capacitor," *International Journal of Control*, vol. 90, pp. 78–89, 2017.

[12] E. Alizadeh, M. Hamzeh, and A. M. Birjandi, "A multifunctional control strategy for oscillatory current sharing in dc microgrids," *IEEE Transactions on Energy Conversion*, vol. 32, no. 2, pp. 560–570, Jun 2017.

Research on the Application of Artificial Intelligence in Long-Distance Pipeline Leakage Monitoring and Risk Prediction

Zeji Li[1]*, Zhanyi Xue[2]

[1]Faculty of Engineering, the University of Hong Kong, Hong Kong 999077

[2]Beijing Ginis Electronics Technology Ltd，Beijing 100085

Abstract: **To address the safety operation issues of oil and gas long-distance pipelines, this study introduces artificial intelligence (AI) technology. First, research on the application of AI in leakage monitoring of long-distance pipelines was conducted. On this basis, the study extended to the application of AI in risk prediction for long-distance pipelines, aiming to lay a foundation for ensuring the safe operation of such pipelines. The findings indicate that, in the process of pipeline leakage monitoring, data collected by various sensors already installed on long-distance pipelines can be utilized to conduct leakage monitoring using machine learning methods. Additionally, under the condition of limited sensor deployment, acoustic monitoring technology can be employed in combination with convolutional neural networks (CNNs) to monitor pipeline leakage. In the process of risk prediction, operational data from pipelines can be collected and feature extraction performed. Based on this, machine learning methods can be used to predict equipment failures along the pipeline. Furthermore, deep learning techniques can be applied to evaluate the risks faced by pipeline operations.**

Keywords: artificial intelligence; long-distance pipeline; leakage monitoring; risk prediction; application research

I. INTRODUCTION

In recent years, with the increasing demand and the continuous expansion of pipeline networks. The long-distance pipeline system has become a key factor for national energy stability and ensuring economic growth. However, the service environment of long-distance oil pipelines is harsh. It needs to meet various geological and climatic conditions. And it also faces issues such as corrosion and third-party damage[1]. Whether it is possible to detect and predict the risk of leaks in long-distance pipelines to reduce the incidence of pipeline leaks and the losses caused by pipeline leaks is a huge challenge. The rapid development of artificial intelligence.

This has brought innovative technological means to pipeline inspection. This provides new ideas and methods for solving the technical problems of leakage monitoring and risk prediction in long-distance pipelines[2]. The goal of this article is to establish a structure based on an intelligent monitoring system. Perform pipeline leakage monitoring and pipeline leakage risk prediction separately. In order to lay the foundation for accurate prediction and intelligent warning of operational risks in pipeline segments.

II. RESEARCH ON THE APPLICATION OF ARTIFICIAL INTELLIGENCE IN LONG-DISTANCE PIPELINE LEAKAGE MONITORING

A. Leakage Monitoring Based on Sensor Data

(1) Sensor Data Collection and Preprocessing

During the process of conducting leak monitoring. For leak detection systems based on sensor data. The acquisition and preprocessing process of sensor data is the guarantee for the entire system to maintain high accuracy and reliability. By analyzing the data information obtained from sensors. Obtain information such as fluid flow rate inside the pipeline. Thus monitoring the operation of the system[3]. During the data collection process. Deploy various types of sensors. Monitor the operation of pipelines or equipment from multiple perspectives. Obtain data from different types of sensors according to predefined sampling frequencies. To obtain data information of a time series. And this is time series data. The system preprocesses the data collected by sensors. The purpose is to ensure that the original observation has a high signal-to-noise ratio. The steps of data preprocessing are shown in Fig.1.

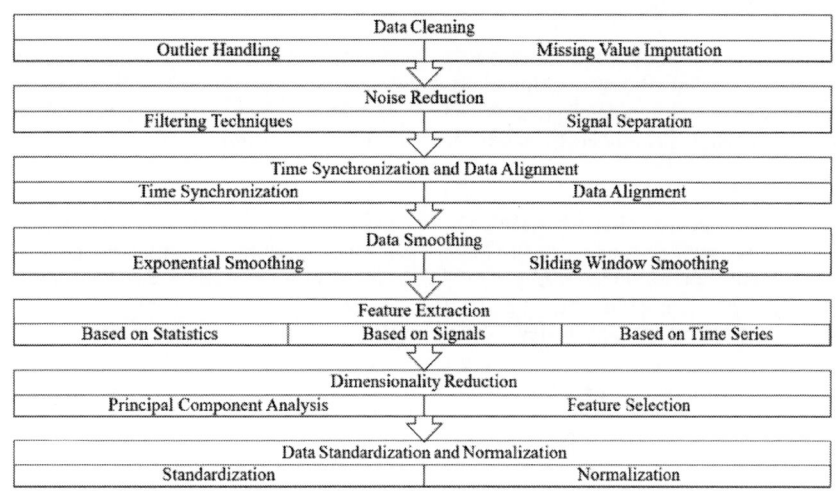

Fig.1 Data Preprocessing Steps Diagram

(2) Feature Extraction and Data Fusion

Feature extraction is a crucial step in extracting leakage related information from multidimensional sensor data. The system extracts time-domain and frequency-domain features from sensor data. Thus constructing multidimensional feature vectors. The common methods for feature extraction and data fusion are shown in Tables 1 and 2.

Table 1 Comparison of Common Data Feature Extraction Methods

Method Category	Method Name	Applicable Scenarios	Advantages	Disadvantages
Based on Statistical Analysis	Mean, Variance, Extreme Value Extraction	Analysis of overall data variation trends	Simple computation; intuitively reflects data fluctuations	Insufficient sensitivity to abrupt signals
	Trend Analysis	Recognizing patterns in data changes over time	Reveals long-term periodic patterns in data changes	Relies on largescale data samples
Based on Signal Analysis	Fast Fourier Transform (FFT)	Periodic signal extraction and frequency domain analysis	Efficiently identifies specific frequency signals caused by leaks	Unsuitable for non-periodic or heavily noisy signals
	Wavelet Transform	Analysis of non-stationary signals	Decomposes both time-domain and frequency-domain features	Complex parameter selection; high computational cost
Based on Time Series Analysis	Time Series Decomposition	Decomposing long-term trends and short-term fluctuations	Distinguishes between stability and anomaly characteristics	Decomposition results depend on data preprocessing
	Moving Window Feature Extraction	Capturing local changes and transient signals	Captures dynamic changes within a small range	Requires determination of an appropriate window size

Table 2 Comparison of Common Data Fusion Methods

Fusion Level	Method Name	Applicable Scenarios	Advantages	Disadvantages
Data-Level Fusion	Data Concatenation	Direct integration of multi-sensor data sources	Simple and intuitive; retains complete raw information	Large data volume; high computational complexity; sensitive to noise
	Weighted Average	Fusion by assigning weights to different data sources	Achieves weighted information integration; reduces the impact of low-quality data	Weight settings rely on experience; difficult to adjust dynamically
Feature-Level Fusion	Feature Concatenation	Combines features from different data sources into a new feature set	Simple to implement; utilizes all feature information	Feature dimensions may become excessively high, increasing computational complexity
	Feature Selection	Selects the most important subset from multiple sources of features	Significant dimensionality reduction; improves model training efficiency	May lose some potentially important information
	Model Weighted Fusion	Assigns weights to different models based on performance	Balances the predictive capabilities of various models	Weight allocation depends on prior knowledge or training processes
	Bayesian Methods	Updates and integrates results from different data sources based on conditional probabilities	Handles data uncertainty; supported by solid theoretical foundation	High algorithmic complexity; heavily reliant on prior distributions

(3) Leak Event Identification Based on Machine Learning

The identification of leakage events in machine learning is the core technical link for achieving intelligent monitoring. It requires the construction of a annotated dataset that includes both normal operating states and various leakage scenarios[4]. Through historical operational data and laboratory simulations of leaks. Establish a comprehensive training database covering different types of leaks and pipeline parameters. Ensure the generalization ability of the model. In terms of algorithm selection. Multiple machine learning methods such as support vector machines and deep learning are often used. Support Vector Machines are suitable for small sample classification problems. Has good generalization performance. Deep neural networks can automatically learn complex spatiotemporal feature patterns. Suitable for processing time-series sensor data. During the model training process. Using cross validation and grid search to optimize hyperparameters[5]. Evaluate model performance through metrics such as precision and recall. To improve real-time performance. We also need to consider the balance between model complexity and computational efficiency. At the final deployment. Establish a multi model integration strategy. Combining voting mechanisms or weighted fusion. Further improve the accuracy and reliability of leak detection. Reduce false alarm rate. Ensure the safe operation of pipelines.

B. Acoustic Monitoring-Based Leak Detection

(1) Acoustic Signal Analysis and Feature Extraction

The analysis of acoustic signals first requires an understanding of the acoustic mechanism of leakage. When a pipeline leaks. High speed fluid is discharged through cracks or holes. Interactions with surrounding media generate turbulence and cavitation phenomena. Stimulate wideband acoustic signals. The signal propagates along the pipeline and surrounding soil. Received by acoustic sensors installed around the pipeline. When analyzing sound wave signals. Firstly, distinguish the leakage signal from other environmental noise, normal fluid flow, and other background signals. Feature extraction includes both time-domain and frequency-domain features[6]. Time domain features include signal amplitude statistical parameters, autocorrelation functions, etc. Reflect the temporal variation characteristics of the leakage signal. Frequency domain features are obtained using fast Fourier transform and power spectral density analysis. For example, main frequency components, frequency center of gravity, etc[7]. The time-frequency characteristics are analyzed using methods such as wavelet transform and short-time Fourier transform. Extract transient characteristics and spectral evolution laws of signals.

(2) Application of Convolutional Neural Networks in Acoustic Signal Analysis

The application of convolutional neural networks in acoustic signal processing can automatically learn the patterns involved. Automatically discover the features within it. Provided an advanced solution for leak detection. The key to applying convolutional neural networks is. How to convert one-dimensional time series acoustic signals into a format suitable for convolutional neural networks to process. For example, converting time-domain signals into stereo spectrogram format. Can be converted into 2D image format[8]. The convenience brought by using convolutional neural networks in image processing with known physical properties. For the application of leak detection. Convolutional neural networks can automatically learn the patterns of leaking acoustic signals. Automatically detect the characteristics of leaked sound signals. And there is no need to manually design features. Design principles. The network will identify spectral features of different types of leaks. The spatial distribution pattern of leaks. Convolutional neural networks are trained on a large amount of annotated data. Can distinguish between noise signals during normal operation and real leakage signals[9].

III. ARTIFICIAL INTELLIGENCE IN RISK PREDICTION FOR LONG-DISTANCE PIPELINES

A. Construction of Risk Prediction Models

(1) Data Collection

The construction of an artificial intelligence long-distance pipeline risk prediction system requires data support from multiple perspectives and sources. Through the process of data collection and organization. Provide data assurance that meets quality requirements for model construction. Data collection is the foundation for the construction of artificial intelligence long-distance pipeline risk prediction models. It is necessary to collect data information from multiple angles and sources in a planned and organized manner. Historical operational data is the main source of data. Including operational parameters such as pipeline pressure and temperature during historical operation, as well as physical property parameters such as pipeline design parameters and environmental factor parameters. Pipeline historical maintenance records and defect repair records[10]. It can provide important support for predicting the occurrence of risks during the historical period of pipelines. Data quality control runs through the entire process of data collection. Establish a data validation matching mechanism. Identify and eliminate abnormal and duplicate data. Regarding the issue of inconsistent data formats among different sources of data information. We need to develop data standard specifications and interface standards.

(2) Feature Selection and Model Training

Feature selection adopts domain expert knowledge, statistical analysis methods, correlation analysis, recursive feature elimination, and other variables closely related to feature selection and risk. Explore the subset of features with the highest contribution to predicting risk. Simultaneously using dimensionality reduction methods such as principal component analysis and linear discriminant analysis. Reduce the dimensionality of the feature space. Avoiding the curse of dimensionality[11]. Due to the combined effects of multiple factors, pipeline risks arise. Therefore, when constructing a pipeline risk prediction model. Train the model using multidimensional features such as pipeline ontology characteristics and historical event characteristics. Support vector machines and random forests are suitable for modeling nonlinear risk relationships. Deep neural networks can mine complex association patterns between features.. The steps of feature selection and model training are shown in Fig.2.

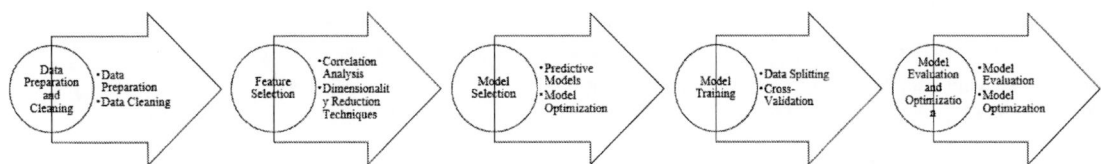

Fig.2 Feature Selection and Model Training Process Diagram

B. Applications of Artificial Intelligence in Risk Prediction

(1) Pipeline Fault Prediction Using Machine Learning Algorithms

Key facilities in pipelines. Such as compressors, valves, oil pumps, etc. Its malfunction will directly affect the safety and stability of the pipeline system. Predicting equipment failures through machine learning algorithms. Thus achieving health management and preventive maintenance of the equipment. Compressors, valves, oil pumps, etc. are important components of pipeline systems. To ensure its normal operation. This study collected operational data for different facilities based on their operational characteristics. The operation data of the compressor mainly collects data such as bearing temperature, speed changes, etc; The valve collects data such as switch position feedback and operation time. Collect data on bearing vibration and pump efficiency of the oil pump. On this basis. Adopting a multi-sensor fusion approach[12]. Build a device health status feature library that integrates mechanical and thermal multiphysics domains. Provide training data basis for machine learning algorithms. There are various types of pipeline facilities. The types of faults also vary. Common faults of compressors include bearing wear, piston ring wear, etc. The main faults of valves are concentrated in actuator faults, locator faults, etc. The main faults of the oil pump include impeller wear, motor failure, etc. This study is based on support vector machine and random forest multi classification fault recognition algorithm. Classify and identify different types of faults. And clarify the key indicator parameters of various faults through feature importance analysis. Deep neural networks have the ability to represent deep nonlinear features. Be able to discover the complex combination patterns of fault characteristic parameters. Thereby improving the accuracy of fault type identification.

(2) Risk Assessment Models Based on Deep Learning

To cope with the different structures and risk characteristics of oil and gas pipelines. Designed multiple deep learning architectures. One dimensional convolutional neural network receives online monitoring data of pressure, temperature and other time series along oil and gas pipelines. Apply the sliding window concept. Explore abnormal patterns and features in local time series along the route. The two-dimensional convolutional neural network receives two-dimensional image data such as ultrasonic images and magnetic flux detection maps generated by pipeline detection. Extract two-dimensional image risk features such as corrosion defects on the inner wall of pipelines and geometric deformations of pipelines. The long short-term memory structure and gated loop unit receive historical data of pipeline operation. Mining long-term dependency information in historical data sequences. Capture the temporal and seasonal patterns of pipeline degradation process. Multi source and multi-scale data for risk detection of oil and gas pipelines include online monitoring data, management data, etc. Multi modal data from different data sources are processed through deep learning models. Apply multimodal fusion technology. Learn the joint distribution of multi-source data. Establish a unified feature representation. Early fusion involves fusing multiple sources of data during the data preprocessing stage. Mid term fusion combines multi-source data at the feature representation level. Late stage fusion combines multimodal prediction results at the decision-making level. The self attention mechanism automatically learns the correlation and relative importance between different data sources. The main risk types of oil and gas pipelines drive the design of specialized prediction sub modules. The corrosion risk prediction module receives inputs such as pipeline material and pipeline protection potential. Combining historical corrosion rate data. Predict the corrosion changes at different locations of the pipeline. The third-party damage risk module receives data such as ground activity monitoring and construction information. Identify high-risk segments for third-party damage. The natural disaster risk module utilizes geological and seismic exploration data as well as weather forecasts. Predict the degree of threat posed by natural disasters such as earthquakes and landslides to pipeline operations. The equipment failure risk prediction module utilizes operational data of key equipment such as compressors and oil pumps. Predict the probability of equipment failure and remaining lifespan. Oil and gas pipelines are a type of linear asset. Its risk has obvious spatial distribution characteristics and temporal evolution laws.

IV. CONCLUSION

The application of artificial intelligence technology in leak monitoring and risk prediction of long-distance pipelines has important theoretical value. The collection and preprocessing of sensor data provide high-quality foundational data for the system. Ensuring the accuracy and reliability of monitoring. Feature extraction and data fusion techniques effectively integrate heterogeneous information from multiple sources. Improved the comprehensive analysis capability of the system. The leakage event recognition method based on machine learning can automatically learn complex fault patterns. Significantly improved detection accuracy and response speed. Meanwhile, the risk assessment model based on deep learning achieves modeling of complex risk relationships through multi-layer neural networks. Improved the accuracy and reliability of predictions.

REFERENCES

[1] Yan X. Application of Artificial Intelligence Technology in Long-Distance Natural Gas Operations[J]. Shanxi Electronic Technology, 2024(04), 14-17.

[2] Li B, Hu J, Zhang Z, et al. Prospects for the Application of Artificial Intelligence in Pressure Pipeline Inspection and Testing[J]. Special Equipment Safety Technology, 2024(04), 18-19.

[3] Liao Q, Liu C, Du J, et al. Application and Prospects of Artificial Intelligence in Oil and Gas Pipeline Operation Management[J]. Oil and Gas Storage and Transportation, 2024, 43(06), 601-613.

[4] Zhang H, Yang H, Wang H, et al. Analysis of the Technical Difficulties in Online Monitoring of Crude Oil Pipeline Leaks in Onshore Oilfields[J]. Henan Science and Technology, 2024, 51(11), 49-52.

[5] Wang Y, Song D, Li B, et al. Research Review on the Use of Acoustic Emission Method for Gas Pipeline Leak Detection[J]. Journal of Safety and Environment, 2024, 24(03), 1114-1127.

[6] Cai C, Yi K, Liao R. Progress in Leak Detection and Location Technology for Long-Distance Oil Pipelines[J]. Science and Technology and Engineering, 2023, 23(24), 10177-10189.

[7] Xiong H. Research on the Construction of an AI-based Oil and Gas Gathering Pipeline Safety Control Platform[J]. Journal of Xi'an Petroleum University (Natural Science Edition), 2023, 38(04), 81-87.

[8] Lu S, Gao K, Jin Y, et al. Research Progress in Oil and Gas Pipeline Leak Detection Technologies[J]. Safety, Health and Environment, 2023, 23(06), 1-10.

[9] Wu X, Li J, Yan J, et al. Progress in Leak Detection and Location Technologies for Subsea Oil and Gas Pipelines[J]. Petroleum Engineering Construction, 2022, 48(03), 1-7.

[10] Yuan M, Gao H, Lu J, et al. Review of Oil and Gas Pipeline Leak Detection Technologies[J]. Journal of Jilin University (Information Science Edition), 2022, 40(02), 159-173.

[11] Gao L, Cao J. Research Progress on Two Difficult Problems in Pipeline Leak Detection[J]. Science and Technology and Engineering, 2021, 21(31), 13203-13210.

[12] Zhang C, Hou N, Lu J, et al. Improved PSO-VMD Algorithm and Its Application in Pipeline Leak Detection[J]. Journal of Jilin University (Information Science Edition), 2021, 39(01), 28-36.

Author Index

An, Zhi	66
Ba, Lei	72
Che, Songtao	39
Chen, Chen	96
Chen, Jing	131
Chen, Lingling	124
Chen, Sichao	111
Chen, Yijing	72
Chu, Wangxu	136
Cui, Yang	96
Dai, Jian	22
Dou, Jie	118
Dou, Zhi	1
Duan, Yuansheng	91
Fang, Kewei	79, 85
Fu, Jinshuo	39
Guo, Wen	58
Guo, Zun	66
Hu, Dalong	58
Hu, Fuquan	101
Hu, Jinfeng	22
Hua, Longsheng	12
Hui, Tianyu	79, 85
Huo, Chao	66
Jiang, Shunjie	58
Kang, Huili	54
Li, Chunhua	72
Li, Guocheng	91
Li, Kai	105
Li, Mengmeng	1
Li, Qianlong	34
Li, Yuzhe	22
Li, Zeji	140
Li, Zheng	72
Liang, Lizheng	1
Lin, Nan	79
Liu, Guang	96
Liu, Tao	79, 85
Liu, Xiaoqiang	101
Lu, Pengfei	29
Lu, Wen	7
Luo, Yuhao	105
Ma, Shengjie	54
Ma, Xiaoxue	91
Ma, Ziluo	49
Meng, Zichao	105
Pan, Hui	111
Peng, Cheng	72
Qian, Yuzhong	1
Qin, Jun	22
Qiu, Haifeng	111
Qiu, Xiangqi	29
Ren, Ran	49
Rong, Xiuting	66
Shen, Lifang	96
Shi, Haining	85
Shi, Hongsi	39
Shi, Lingzhen	111
Shi, Wusheng	58
Song, Jilong	39

Su, Lei	22
Tang, Tang	79
Tao, Can	91
Tao, Xin	1
Wang, Binfeng	79
Wang, Jiaxing	66
Wang, Kai	39
Wang, Qiang	79, 85
Wang, Tao	72
Wang, Xiaowei	101
Wang, Xiaoyu	7
Wang, Zongliang	136
Wei, Song	29
Wen, Jiaming	136
Weng, Lei	54
Weng, Liguo	111
Wu, Han	72
Wu, Junkai	17
Wu, Weixun	136
Wu, Zhaoyuan	66
Xue, Zhanyi	140
Yan, Shubin	96
Yan, Wensheng	54
Yang, Lin	101
Yang, Minjing	105
Yang, Yun	105
Yang, Yuxuan	66
Ye, Hengzhi	22
Ying, Hong	79, 85
Zhang, Junyang	72
Zhang, Tao	85
Zhang, Wenhai	22
Zhang, Wenying	49
Zhang, Xiao	29
Zhang, Zhen	54
Zhang, Zhihao	136
Zhao, Yue	105
Zhi, Chaoyang	44
Zhou, Ming	66
Zhou, Yuxin	58
Zhu, Jianquan	105
Zhu, Zhenqing	1

IEEE
445 Hoes Lane
Piscataway, NJ 08854-4141

ISBN 979-8-3315-2490-6